TWENTY FIRST CENTURY

science

GCSE Chemistry

Nuffield Curriculum Centre

THE UNIVERSITY of York

OXFORD

The exercises in this Workbook cover the OCR requirements for each module. If you do them during the course, your completed Workbook will help you revise for exams.

Project Directors
Jenifer Burden
John Holman
Andrew Hunt
Robin Millar

Project Officers
Peter Campbell
Angela Hall
John Lazonby
Peter Nicolson

Course Editors
Jenifer Burden
Peter Campbell
Andrew Hunt
Robin Millar

Authors
Andrew Hunt
John Lazonby
Caroline Shearer

Contents

WORKBOOK

Great Clarendon Street, Oxford OX2 6DP

Oxford University Press is a department of the University of Oxford.
It furthers the University's objective of excellence in research, scholarship,
and education by publishing worldwide in

Oxford New York

Auckland Cape Town Dar es Salaam Hong Kong Karachi
Kuala Lumpur Madrid Melbourne Mexico City Nairobi
New Delhi Shanghai Taipei Toronto

With offices in

Argentina Austria Brazil Chile Czech Republic France Greece
Guatemala Hungary Italy Japan Poland Portugal Singapore
South Korea Switzerland Thailand Turkey Ukraine Vietnam

First published 2006

British Library Cataloguing in Publication Data

Data available

ISBN: 978-0-19-915054-0

10 9 8 7 6 5

Printed by Bell and Bain Ltd., Glasgow

Illustrations by IFA Design, Plymouth, UK

These resources have been developed to support teachers and students undertaking a new OCR suite of GCSE Science specifications, Twenty First Century Science.

Many people from schools, colleges, universities, industry, and the professions have contributed to the production of these resources. The feedback from over 75 Pilot Centres was invaluable. It led to significant changes to the course specifications, and to the supporting resources for teaching and learning.

The University of York Science Education Group (UYSEG) and Nuffield Curriculum Centre worked in partnership with an OCR team led by Mary Whitehouse, Elizabeth Herbert and Emily Clare to create the specifications, which have their origins in the Beyond 2000 report (Millar & Osborne, 1998) and subsequent Key Stage 4 development work undertaken by UYSEG and the Nuffield Curriculum Centre for QCA. Bryan Milner and Michael Reiss also contributed to this work, which is reported in: 21st Century Science GCSE Pilot Development: Final Report (UYSEG, March 2002).

Sponsors
The development of Twenty First Century Science was made possible by generous support from:
- The Nuffield Foundation
- The Salters' Institute
- The Wellcome Trust

wellcometrust

Air quality - Higher

1 The Earth is surrounded by a thin layer of atmosphere made up of a mixture of gases.

The pie chart below shows the percentages of different gases that make up the Earth's atmosphere. Complete the key for the pie chart.

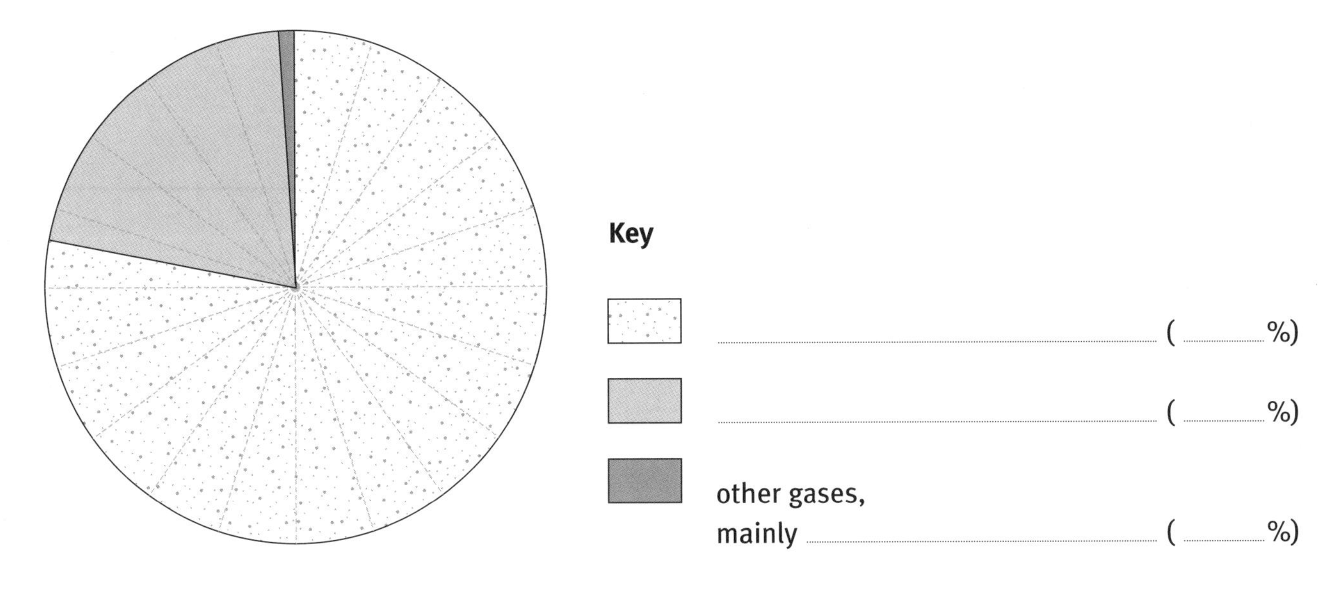

2 Human activity adds to the air small amounts of extra chemicals called atmospheric pollutants.

Complete the table below to show what problem each pollutant causes.

Name of pollutant	Problem(s) caused
sulfur dioxide (S, O, O)	
carbon monoxide (C, O)	
nitrogen dioxide (O, N, O)	
particulates (tiny bits of solid suspended in the air)	

3 A set of measurements of an atmospheric pollutant can be used to obtain a best estimate of the true value of the concentration of the pollutant.

Imagine you and other students in your class have measured the concentration of sulfur dioxide in the air near a factory. You used the same instrument and all the measurements were completed within a short space of time. The table shows the measurements.

	Sulfur dioxide concentration (parts per billion)		
Student	**One**	**Two**	**Three**
A	16.1	15.9	16.0
B	15.9	16.0	16.1
C	16.1	17.3	16.1
D	16.3	16.3	16.2
E	16.2	15.9	15.9

a Give one possible reason why all of the repeated measurements are not exactly the same.

....

....

You want to obtain a measurement that is as accurate as possible. That is a measurement that is as close as possible to the true value of the concentration of sulfur dioxide. Use the steps **b** to **e** to do this.

b Check the measurements to see if any of them are outliers. These are measurements that are so different from the others that they are likely to be wrong.

- Plot all of the measurements along this axis. Use a small cross for each measurement.
- Draw a circle around any that you think are outliers.

c Outliers are not used. Calculate the mean value of the remaining measurements.

....

....

17.3
17.2
17.1
17.0
16.9
16.8
16.7
16.6
16.5
16.4
16.3
16.2
16.1
16.0
15.9

d The mean value is used as your best estimate of the concentration of sulfur dioxide.

The best estimate = ... parts per billion

e You can't be absolutely sure that your best estimate is the true value, but you know that it must be somewhere between the lowest and highest measurement you have used. This is called the range.

The range of values is:

from .... to .... parts per billion.

There is not a real difference between two measurements of the concentration of a pollutant if the ranges of the measurements overlap.

f You then measured the concentration of sulfur dioxide outside your school.

Explain why, if there is a real difference between the concentration of sulfur dioxide outside the factory and outside your school, the ranges of the two sets of readings will not overlap.

	Sulfur dioxide concentration (parts per billion)		
Student	**One**	**Two**	**Three**
A	15.4	15.4	15.2
B	15.6	15.5	15.4
C	15.5	15.5	15.4
D	15.3	15.3	14.1
E	15.6	15.5	15.4

....

....

....

Use the data in the table to check if there is a real difference. Explain your answer.

....

....

....

g Now go back through this task and underline or highlight the first time each of these key words is met and make sure you understand what it means.

repeated measurements	accurate	outlier	mean	best estimate	range	real difference

h If your friend had wanted to take only one measurement rather than several measurements, how would you explain that it is better to use repeated measurements?

....

....

....

....

4 When fuels burn, they combine with oxygen from the air. This involves atoms being rearranged.

The diagram below compares what goes into a car engine with what comes out.

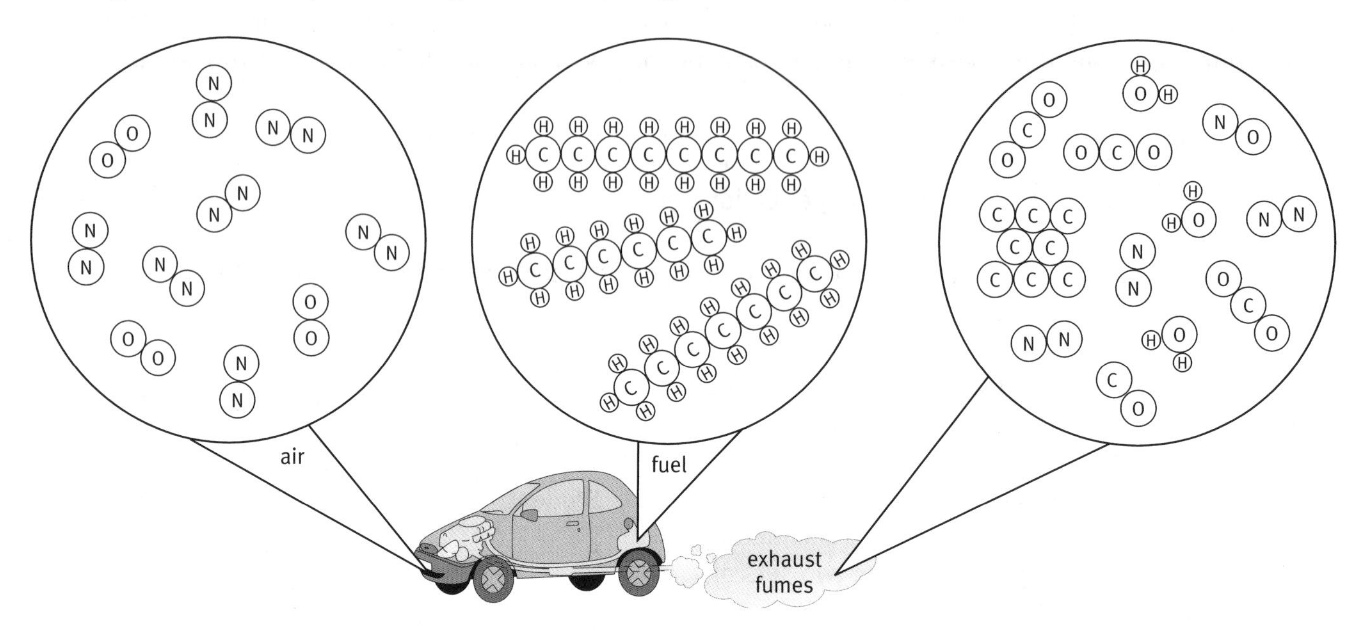

When drawing pictures to represent atoms it is usual to use particular colours for particular atoms:

nitrogen = blue	oxygen = red	carbon = black	hydrogen = white

a Air and fuel (petrol or diesel) go into a car's engine.
Air is mostly a mixture of nitrogen and oxygen.
Fuel is a mixture of hydrocarbons (chemicals made up of hydrogen and carbon).

➲ Colour in all the atoms in the picture of air and in the picture of fuel.

b One of the chemicals that comes out of the engine is nitrogen monoxide (N)(O).

You can work out how it has been formed by comparing the pictures of what goes in and what comes out of the engine.

➲ Colour the atoms in nitrogen monoxide in the exhaust fumes.

➲ Then colour all the atoms in the following explanation.

Nitrogen monoxide (N)(O) must have been formed by the atoms in (O)(O) and (N)(N) separating to form (O) and (N) atoms, which then combine with each other to form (N)(O).

c The percentage of NO formed is very small. Most of the exhaust fumes is made up of carbon dioxide (CO_2), water vapour (H_2O), and unchanged N_2 from the air.

➲ Look for these chemicals in the exhaust fumes and colour them in.

d Use ideas of atoms separating and then joining together in different ways to explain how the CO_2 and H_2O must have been formed.

e Look at the pictures on page 6.

➔ Find two other pollutants in the exhaust fumes. Colour their atoms.

These diagrams compare what goes into a coal-fuelled power station and gas-fuelled power station with the main chemicals that come out of the chimney.

f Use words and/or pictures to explain how CO_2 is formed in a coal-fuelled power station and in a gas-fuelled power station.

Coal:

Gas:

g In addition to the main gases shown above, power stations can produce small amounts of atmospheric pollutants. Write the names of four air pollutants that can be formed and a brief explanation of how each is formed.

Name:

How it is formed:

Name:

How it is formed:

Name:

How it is formed:

Name:

How it is formed:

5 The products of burning depend on the fuel and the amount of oxygen available.

a Complete the tables below to show the products made when different fossil fuels are burned.

Main products

Fossil fuel	Made up of . . .	Products of complete burning
coal	mainly carbon	....
gas	methane, which is the hydrocarbon CH_4	.... and .... (hydrogen oxide)
petrol, diesel, fuel oil	mixtures of hydrocarbons	.... and .... (hydrogen oxide)

Minor products if oxygen supply is limited

Fossil fuel	Made up of . . .	Minor product formed on burning in limited oxygen
coal	mainly carbon	.... and tiny pieces of carbon, called ....
gas, petrol, diesel, fuel oil	hydrocarbons	.... and tiny pieces of carbon, called ....

Minor product also formed if sulfur is present in fuel

Fossil fuel	Made up of . . .	Minor product formed on burning
coal	mainly carbon plus small amount of sulfur	....
gas, petrol, diesel, fuel oil	hydrocarbons plus small amount of sulfur*	....

* this sulfur can be removed before the fuel is burned.

b Complete this sentence:

Air is made up of a mixture of nitrogen and oxygen. When it is heated to a very high temperature in a furnace or an engine, the minor product formed is ....

6 You need to be able to recognize diagrams of molecules and formulae of some chemicals.

Draw a line to link each diagram to the correct formula. Then name the chemical. One example has been done for you.

Chemical model	Formulae	Chemical name
O S O	CO	
C O	CH_4	
O N O	SO_2	sulfur dioxide
H O H	NO_2	
O C O	H_2O	
H H C H H	NO	
N O	CO_2	

7 **You need to be able to recognize diagrams showing the rearrangement of atoms during some chemical reactions.**

Draw a line to link each diagram to the description of the reaction it represents.
One has been done for you.

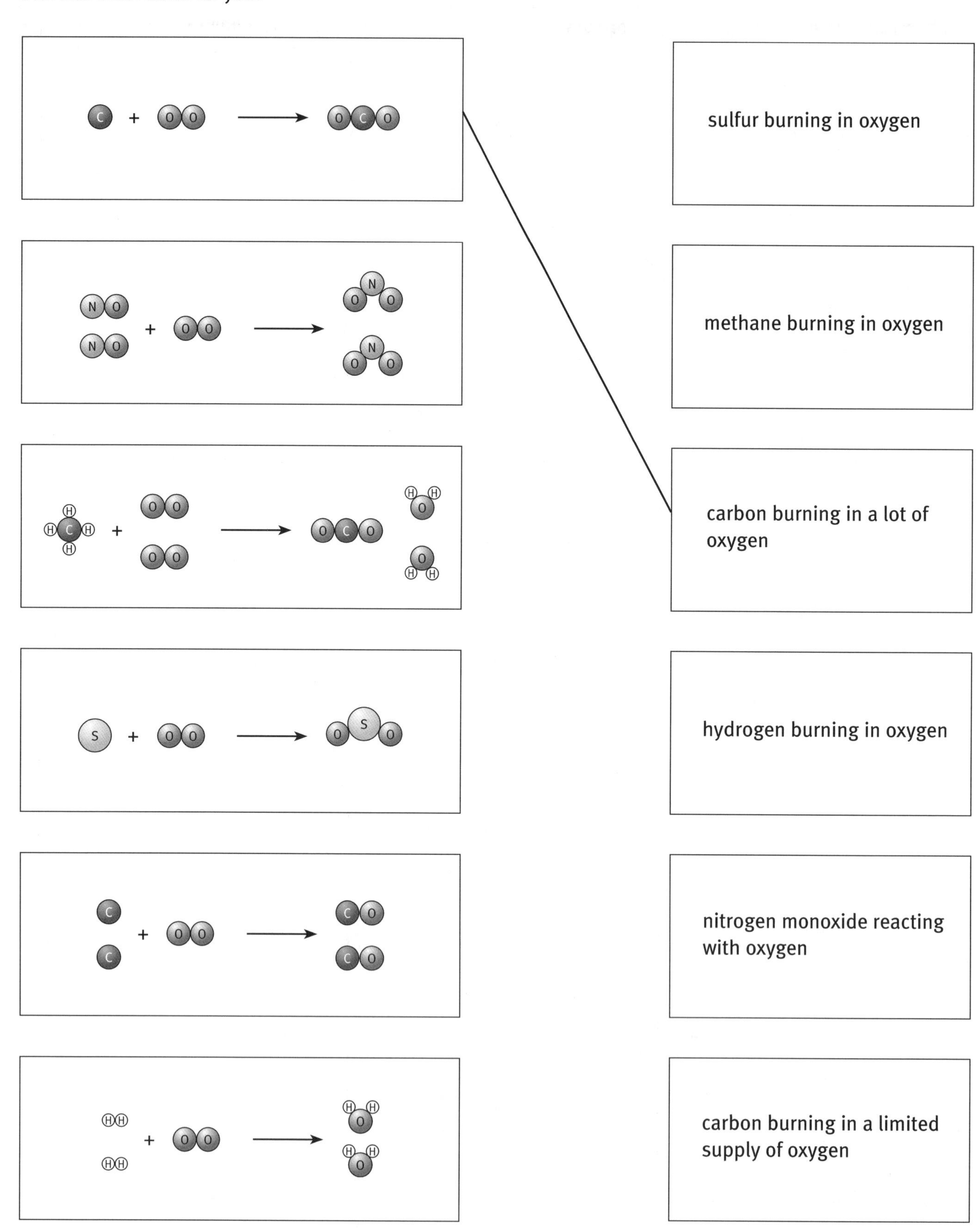

8 In a chemical reaction, the properties of the reactants and products are different.

Choose reactions from page 10 to fit these descriptions. (All the reactions are at room temperature.) The first one has been done for you.

a The reactants are gases and the product is liquid.

hydrogen burning in oxygen

b One reactant is a black solid and the product is a very poisonous gas.

c All the reactants and products are gases.

d The reactants are a yellow solid and a colourless gas, and the product dissolves in water to form an acid.

e The reactants are a black solid and a colourless gas, and the product is a colourless gas that turns limewater milky.

f Two gases react together to form a liquid that turns anhydrous copper sulfate blue and a gas that turns limewater milky.

9 In a chemical reaction, the numbers of atoms of each element must be the same in the products as in the reactants.

a Methane burns in oxygen to form carbon dioxide and water. The chemical reaction can be represented by:

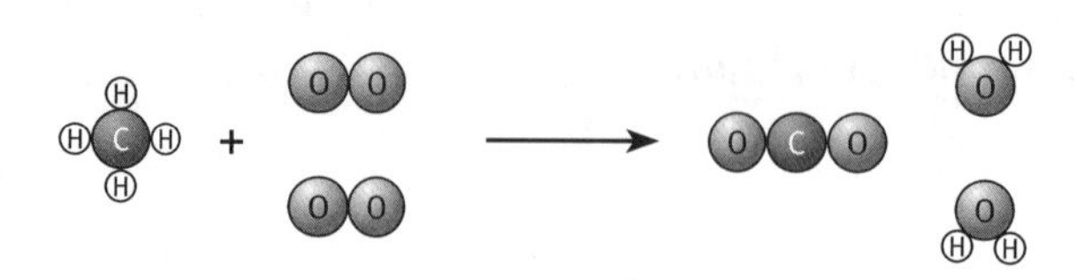

Number of atoms in reactants	C =
	O =
	H =

Number of atoms in products	C =
	O =
	H =

Complete the boxes above. Fill in the number of each kind of atom in the reactants and in the products.

When a chemical reaction such as burning occurs, no atoms are destroyed and no new ones are made. The atoms are **conserved**.

b When you have a barbecue, you burn the fuel to produce the heat needed to cook your food. When you have finished you have used up some fuel. But you have not used up the atoms in the fuel – the atoms are still somewhere. They have been conserved.

- If you used charcoal as the fuel, what happens to the atoms in the charcoal?

....

....

- If you have used a gas barbecue, the fuel is natural gas (methane CH_4) or butane (C_4H_{10}). What happens to the atoms in the fuel?

....

....

c Explain why, from the point of view of air quality, it is better to 'get rid' of garden rubbish by composting rather than by burning?

....

....

....

10 Some atmospheric pollutants are removed from the atmosphere, some stay in the atmosphere for a long time, and some react with other chemicals to form different pollutants.

a Complete the spider diagram to show what happens to some atmospheric pollutants.

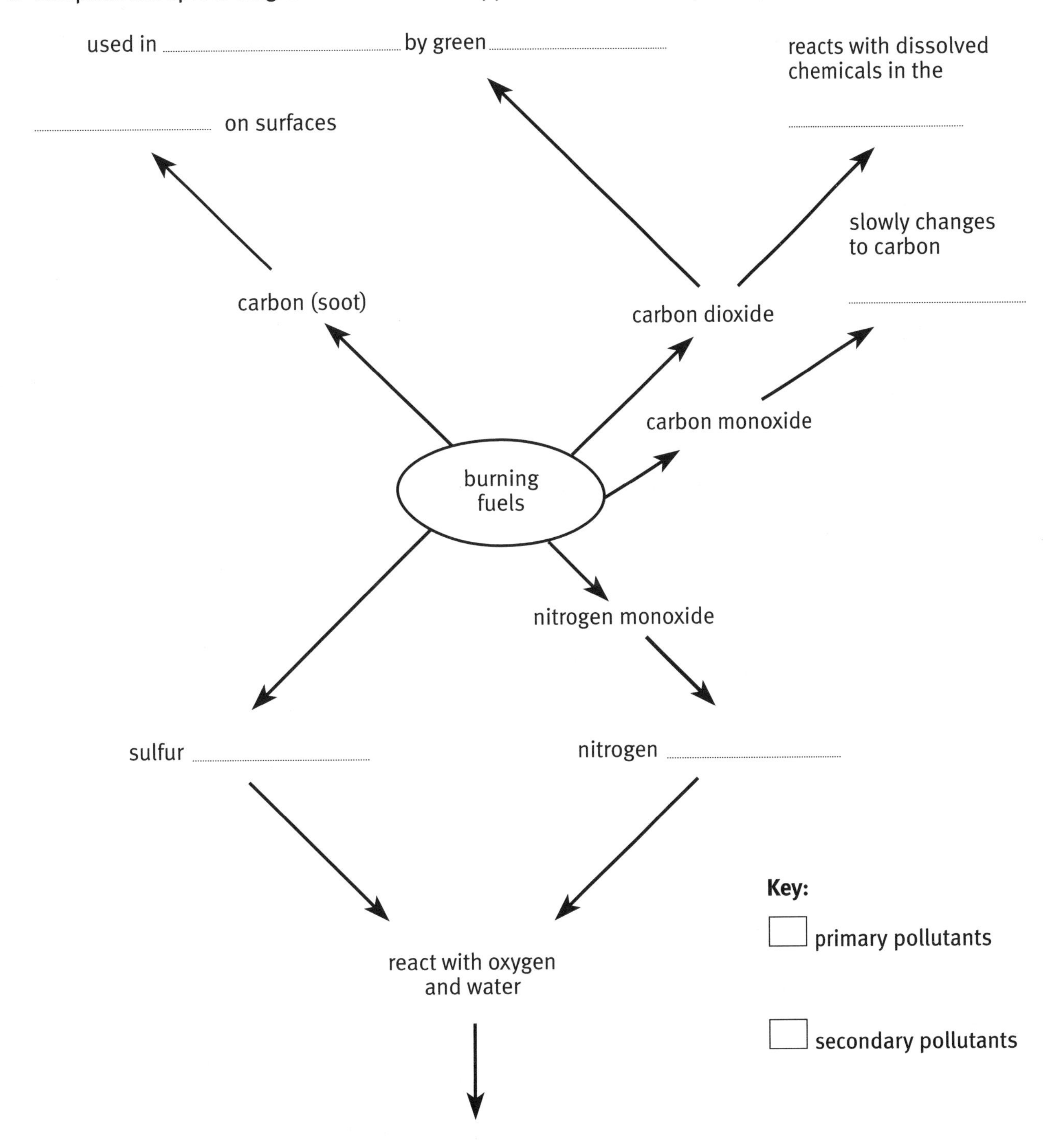

b On the diagram above, highlight or underline the **primary pollutants** with one colour and the **secondary pollutants** with a different colour. Colour the key.

c Explain why it is difficult to decide whether CO_2 should be called a pollutant.

....

....

11 A correlation is a link between two things. For example, as one thing increases the other also increases.

a Graph **A** shows how the pollen count changes during a typical year.

Complete this description of what the graph shows. Put a ring around the correct **bold** words.

Between the months of May and July the pollen count **increases** / **decreases**.

Between the months of August and December the pollen count **increases** / **decreases**.

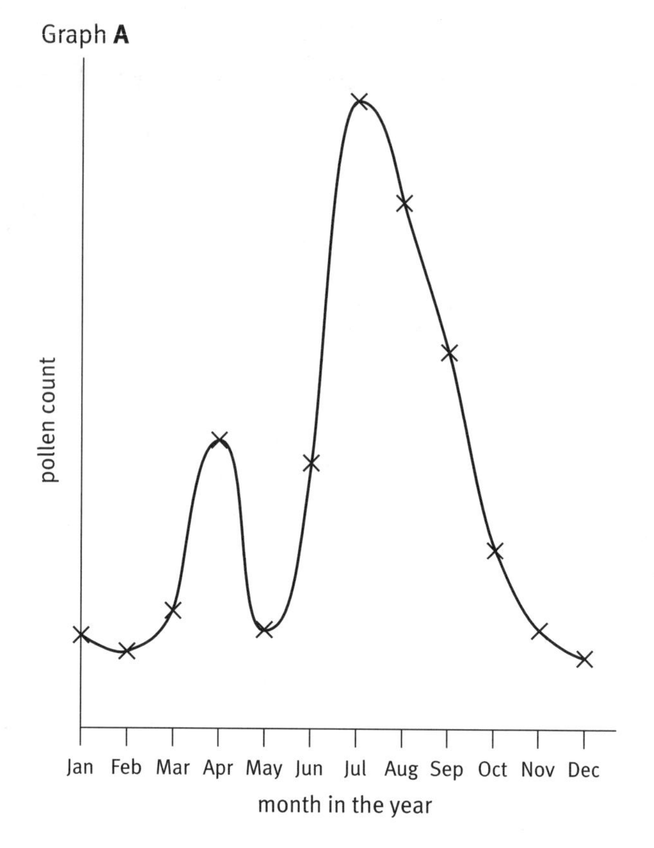

b Graph **B** shows how the number of people with hay fever symptoms changes during a year.

Compare graphs **A** and **B** by completing the following sentences.

As the pollen count increases the number of people with hay fever symptoms

....

As the pollen count decreases the number of people with hay fever symptoms

....

The graphs show that there is a correlation between pollen count and the number of people with hay fever symptoms.

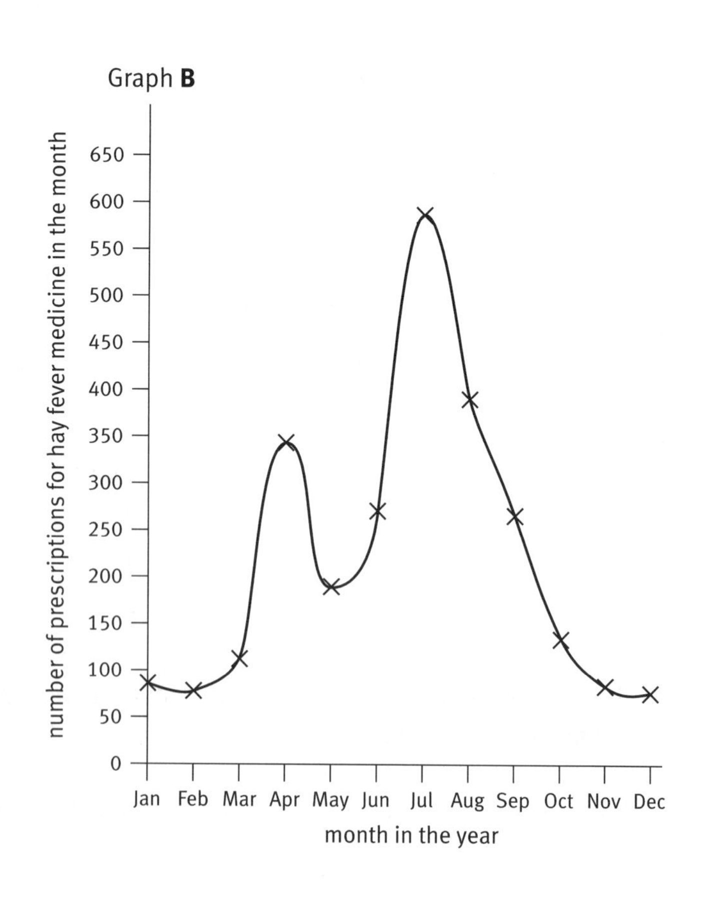

12 A correlation doesn't always mean that one thing increasing causes the other to increase.

a In the last question you saw that there is a correlation between the pollen count and the number of people who suffer from hay fever symptoms.

There may be other things that can be measured which will correlate with hay fever symptoms.

Put a tick or a cross next to each of the following to show which of the factors you would expect to correlate, in the same way as pollen count does, with hay fever symptoms. Two have been done for you.

- ice cream sales ✔
- the size of electricity bills for your home ✘
- hours of sunshine per day
- number of people away on holiday
- depth of the snow on Alpine ski slopes

b There are three types of reason why a factor (such as pollen count) will correlate with an outcome (such as hay fever):

(1) pollen causes hay fever

(2) another factor causes both the pollen count and the hay fever symptoms to increase

(3) it is just a coincidence

In the case of pollen and hay fever it is reason number (1) because there is evidence that hay fever is an allergic reaction to pollen. An increase in the pollen count does cause an increase in the hay fever symptoms.

Explain which of reasons (1), (2), or (3) best explains why ice cream sales correlate with hay fever symptoms.

....

....

....

13 Improvements to motor vehicle and power station technology can lead to reductions in the amount of pollutants produced.

a Some of the most important technological developments which aim to improve air quality are listed in the left-hand column.

Draw a line to match each development to the correct explanation of what it does. One has been done for you.

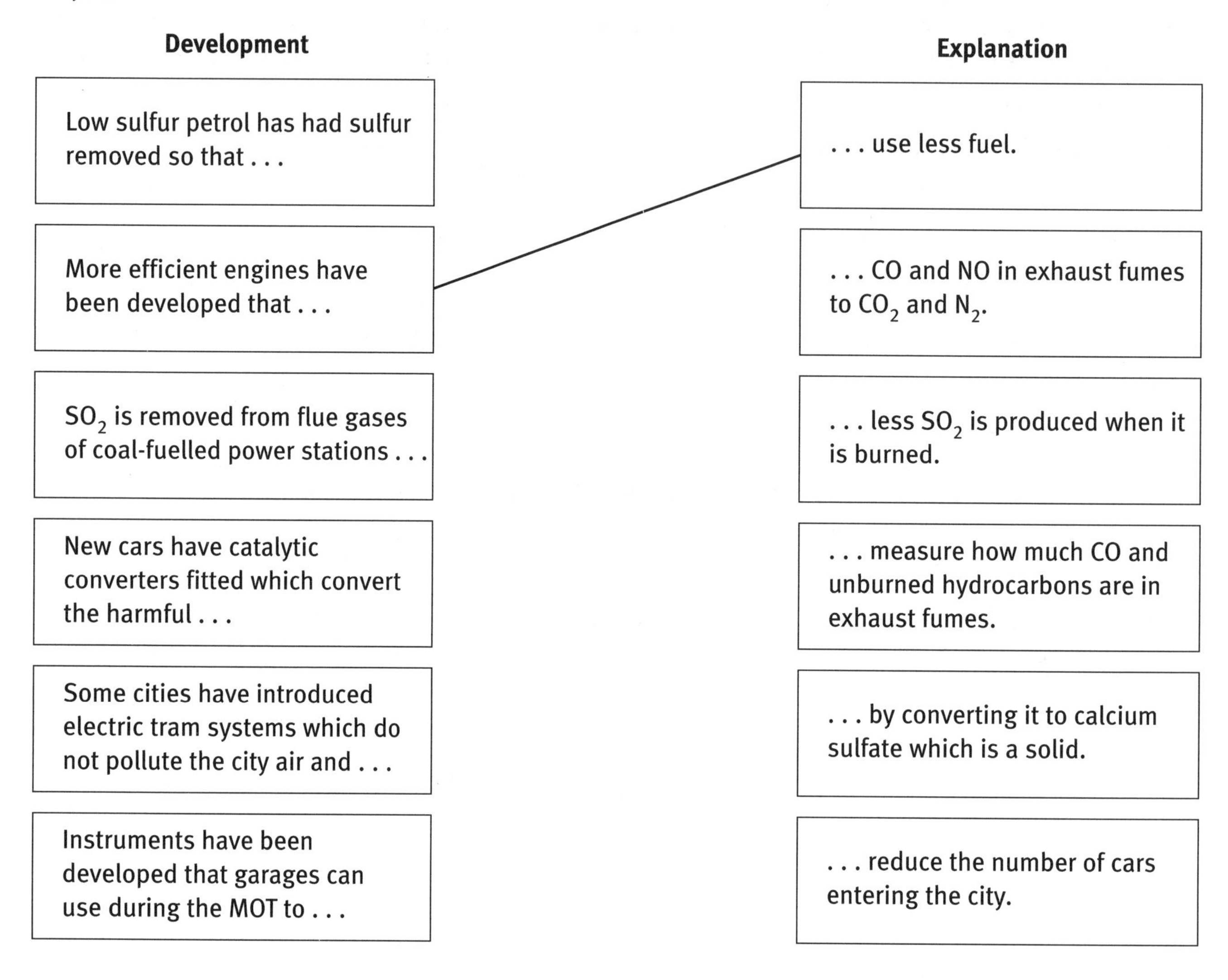
Development
Explanation
Low sulfur petrol has had sulfur removed so that . . .
More efficient engines have been developed that . . .
SO_2 is removed from flue gases of coal-fuelled power stations . . .
New cars have catalytic converters fitted which convert the harmful . . .
Some cities have introduced electric tram systems which do not pollute the city air and . . .
Instruments have been developed that garages can use during the MOT to . . .
. . . use less fuel.
. . . CO and NO in exhaust fumes to CO_2 and N_2.
. . . less SO_2 is produced when it is burned.
. . . measure how much CO and unburned hydrocarbons are in exhaust fumes.
. . . by converting it to calcium sulfate which is a solid.
. . . reduce the number of cars entering the city.

b Fill in the missing words to complete these additional notes on two of the developments.

- The SO_2 in flue gases from coal power stations is reacted with a mixture of air, water, and ....
- The catalyst in a catalytic converter makes the chemical reactions happen more ....

14 Government laws and regulations, financial incentives, and individual actions can reduce atmospheric pollution.

a Draw a line or lines linking each of the ways of improving air quality to the type(s) of action that best describe it.

Ways of improving air quality	Types of action
Fit catalytic converters to motor vehicles.	
Increase the car tax on cars with big engines.	Government actions
Use MOT tests to enforce limits on vehicle emissions.	
Remove SO_2 and particulates from power station flue gases.	
Make international agreements to reduce CO_2 emissions.	financial incentives
Stop leaving TVs on standby.	
Provide grants to improve house insulation.	technological developments
Make fewer car journeys.	
Develop and use more efficient engines.	
Remove sulfur from natural gas and fuel oil.	individual actions
Use low energy light bulbs.	

b In the empty box, add another way of improving air quality that you think is worth including. Link it up to the type(s) of action that best describe it.

Material choices – Higher

C2

1 Different materials for different jobs

The diagram shows objects made of different materials, and some different properties of the materials. Draw lines to match each property with one material. One has been done for you.

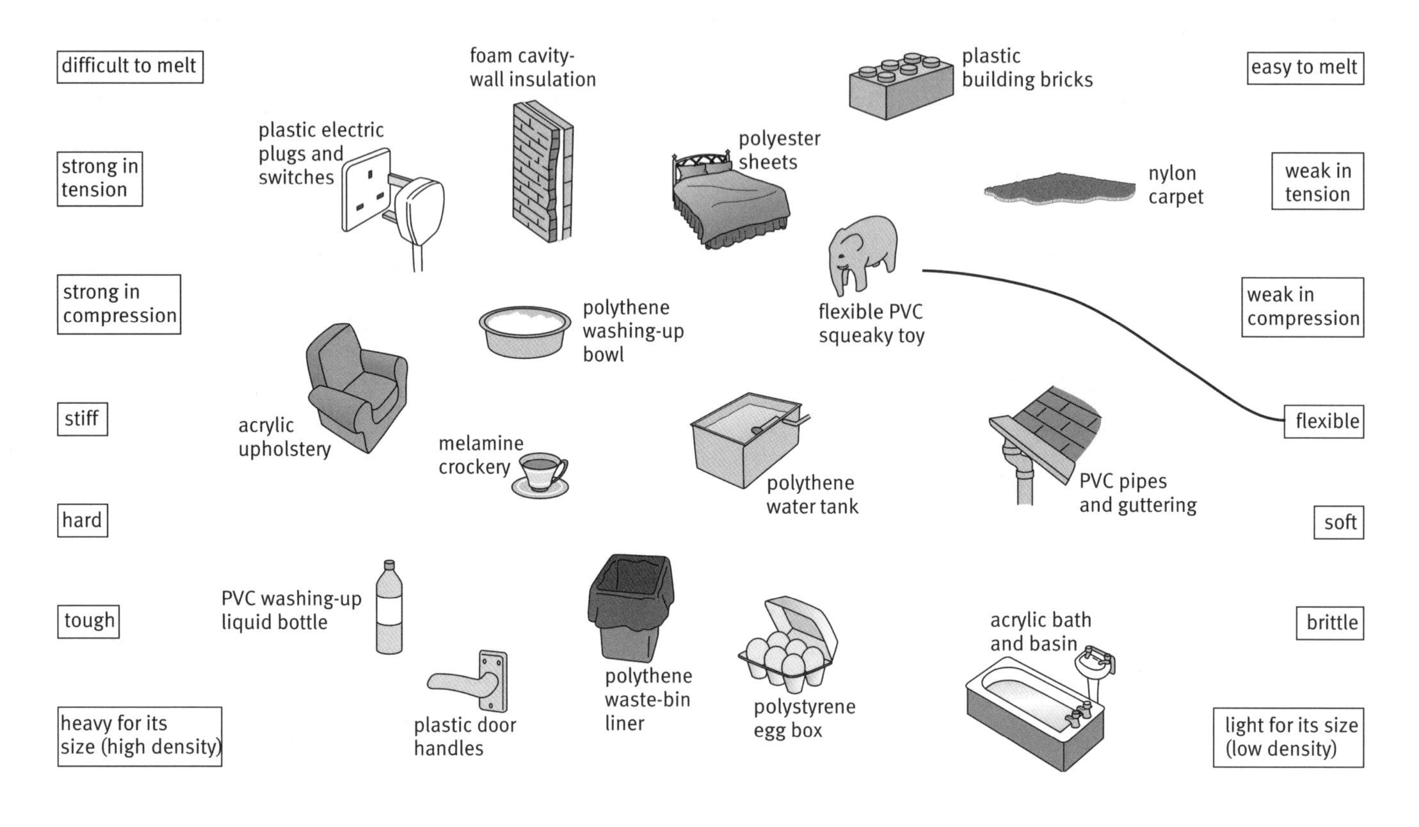

2 Properties of materials

Draw lines to link each property name to its definition.

Property name	Definition
strength in tension	how heavy the material is for its size
strength in compression	how hard it is to bend the material
stiffness	how hot the material has to be before it turns liquid
hardness	how much squashing force is needed to break the material
melting point	how much stretching force is needed to break the material
density	how easy it is to scratch or dent the material

3 Choosing the right material

The table shows some of the properties for four polymer materials.
(A polymer is a large molecule made of lots of small ones joined together.)

Polymer	Strength in tension (MN/m^2)	Stiffness (GPa)	Density (g/cm^3)	Highest working temperature (°C)
A	60	4.1	1.39	60
B	80	0.7	1.12	120
C	15	0.2	0.92	85
D	40	3.0	1.05	65

Write the letter of the polymer beside each description. Then give a reason for your choice.

a Has strength and high stiffness: ....

Reason for this choice: ....

....

b The most flexible: ....

Reason for this choice: ....

....

c The best to make a bowl to contain boiling water: ....

Reason for this choice: ....

....

d The best to make fishing line: ....

Reason for this choice: ....

....

e The best to make guttering for a house: ....

Reason for this choice: ....

....

f The best to make wrapping film for food: ....

Reason for this choice: ....

....

4 Testing materials

A student tested carrier bags to find out how easily they would rip. She compared bags from supermarkets A and B. She used this equipment to test 20 mm × 100 mm strips from each type of bag. The load on each strip was increased by adding 20 g masses until the strip broke.

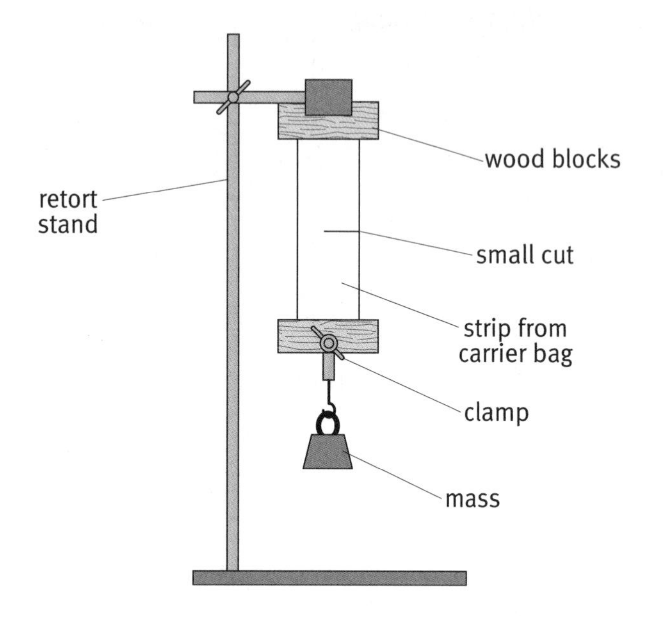

Load needed to break strips from type A bags (g)	Load needed to break strips from type B bags (g)
360	760
380	720
360	780
340	740
340	720
380	860
360	780
380	740
340	760
340	760

a Should all the samples for one supermarket be cut from one bag or from different bags? Explain why you think this.

....

....

....

b Give a possible reason why all ten measurements for the bags from supermarket **A** were not the same.

....

....

c Explain why using ten samples from each type of bag can make the results more accurate.

....

....

d Work out the mean value for the bags from supermarket **A.** Show your working.

Answer ....

e What is the range of values for the bags from supermarket **A**?

The range is from .... to ....

f Mark the ten readings for type **B** bags on this number line. Draw a cross for each value.

700 720 740 760 780 800 820 840 860

Draw a circle round any values that are outliers.

Suggest two reasons which might explain why there is an outlier in the results.

- ....

....

- ....

....

g Now work out the mean value for the bags from supermarket **B**. Do not include outliers.
Show your working.

Answer ....

h What is the range of values for the bags from supermarket B?

The range is from .... to ....

i Compare the mean and range for the two types of bag.

....

....

j What conclusions can you come to from the results?

....

....

5 Comparing materials

The table shows the results of tests on six different types of screwdriver. The results compare the strength of the screwdrivers when used with screws. The results also show how well the screwdrivers stood up to being hit with a hammer.

Type	Price (£)	Handle	Strength as screwdriver	Use with a hammer
A	1.11	plastic	☆	☆☆
B	3.76	plastic	☆☆☆	☆☆☆
C	4.25	plastic	☆☆☆☆☆	☆☆☆☆
D	7.95	wood	☆☆☆	☆☆
E	3.50	wood	☆☆☆	☆☆☆
F	1.99	wood	☆☆	☆☆

a Which type of screwdriver is the cheapest? ....

b Which type is the strongest? ....

c The testers checked five of each type of screwdriver. Explain how this makes the results better.

....

....

d Suggest two ways in which the testers could make sure that they can fairly compare the results for each type of screwdriver.

- ....
- ....

e Suggest two reasons why not all the plastic handles gave the same results in the hammer test.

- ....
- ....

f Do the results in the table show that plastic handles are better than wooden handles? Give your reasons.

....

....

g Using the information in the table, which screwdriver would you buy? Give your reasons.

....

....

....

6 Polymers

a Use these words to complete the paragraph.

joined	chains	polymerization	polymers	polymer
monomers	long	poly	blocks	

Polymers are made up of very .... molecules. This is true for natural .... such as silk or cotton, and synthetic polymers such as polythene or nylon. They are .... of smaller molecules. .... means many, *mer* means part, so .... means many parts. The building .... that make the polymer are called .... (*mono* means one). So lots of .... up monomers make one polymer. The process of making a polymer is called ....

b Now use words from the paragraph to help you label this diagram.

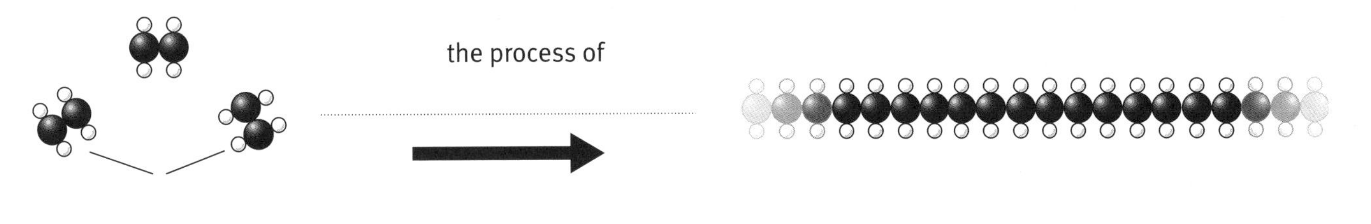

7 Forces between molecules

Both of the diagrams below show hydrocarbon molecules. Diagram **A** shows methane. Diagram **B** shows polythene. Complete the labels for each diagram. Use these words and phrases.

gas	hydrocarbon	molecules	little	more	chain polymer	weak	stronger

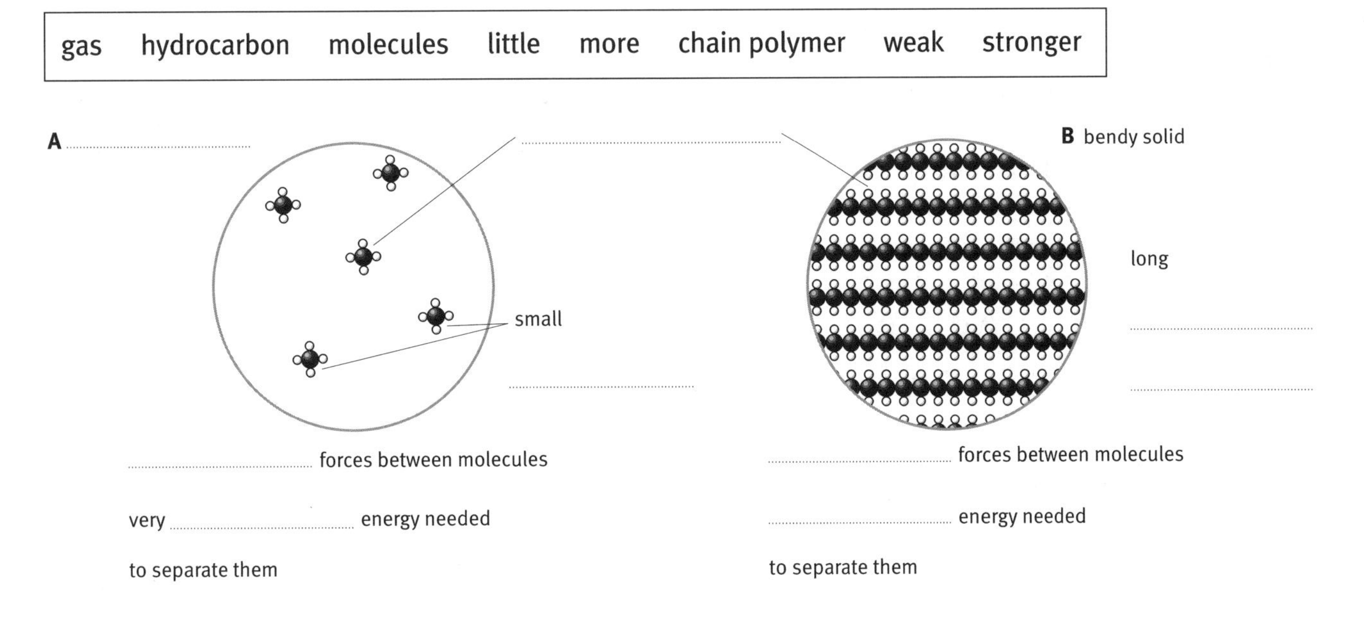

8 Changing a polymer by changing its molecules

In the diagrams below, a polymer molecule such as:

is represented by a line

Two types of polythene

These diagrams show the molecules in two types of polythene: low-density polythene (**ldpe**) and high-density polythene (**hdpe**).

a Label the diagrams to show which is which. Label the diagram that shows a crystalline material.

.... -density polythene .... -density polythene

b Complete this sentence.

High-density polythene is a little bit denser than low-density polythene because

....

Natural and synthetic rubber

These diagrams show the molecules in natural rubber and vulcanized rubber.

c Label the diagrams to show which is which. Label a cross-link.

.... rubber .... rubber

d Complete this sentence.

It takes very little force to break natural rubber because ....

....

Two types of PVC

These diagrams show molecules in unplasticized PVC (uPVC) and plasticized PVC.

e Label the diagrams to show which is which. Label a plasticizer molecule.

.... PVC .... PVC

f Complete this sentence.

Compared with uPVC, plasticized PVC is ....

9 Natural and synthetic materials

a Some materials are natural – they may come from plants or from animals. Others are manufactured – they are called synthetic materials.

Sort these eleven materials into the table below.

acrylic	cotton	leather	nylon	paper	polyester
polythene	PVC	silk	wood	wool	

Natural materials		**Synthetic materials**
From plants	**From animals**	

b A hundred years ago most materials in a kitchen were made from wood, natural fibres, metal, or pottery (ceramic).

The picture shows an old kitchen.
On the picture:

- use one colour to shade in objects that might today be made of synthetic polymers
- use a different colour to shade in objects that could not be made of synthetic polymer
- fill in the key with the colours you have used.

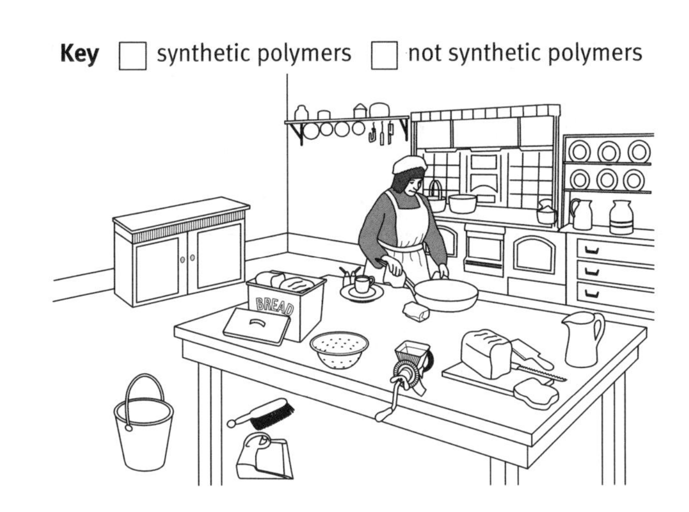

10 Changing chemicals from crude oil into synthetic materials

Match the beginning and end of each sentence. One has been done for you.

Crude oil found in the Earth's crust is ...	... carbon and hydrogen atoms.
Crude oil consists mainly of chemicals called ...	... a synthetic material.
Hydrocarbons are chemicals made from ...	... fuels, lubricants, and chemicals.
Petrochemical refineries produce ...	... hydrocarbons.
A polymer made of chemicals from oil is ...	... a dark, tarry liquid.

11 The life of manufactured products

a Use these words to complete the statements for the three main stages in the life of a product.

cleaning	dispose	maintain	manufacture	petrol	raw materials	space

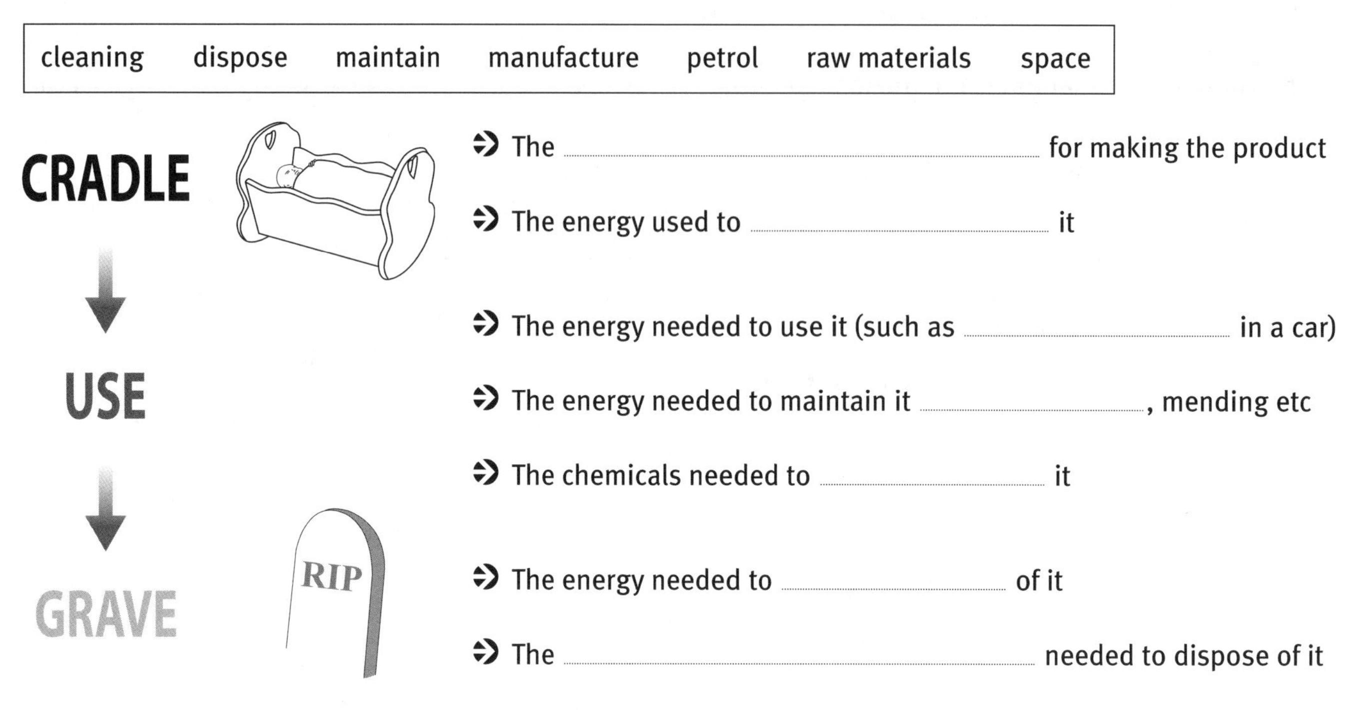

- The .... for making the product
- The energy used to .... it
- The energy needed to use it (such as .... in a car)
- The energy needed to maintain it ...., mending etc
- The chemicals needed to .... it
- The energy needed to .... of it
- The .... needed to dispose of it

b Use these words to label the life cycle of a synthetic polymer.

chemicals	crude oil	electricity	energy recovery	fuels	incineration	landfill	oil well
plastic bottles	recycling	refinery	reuse				

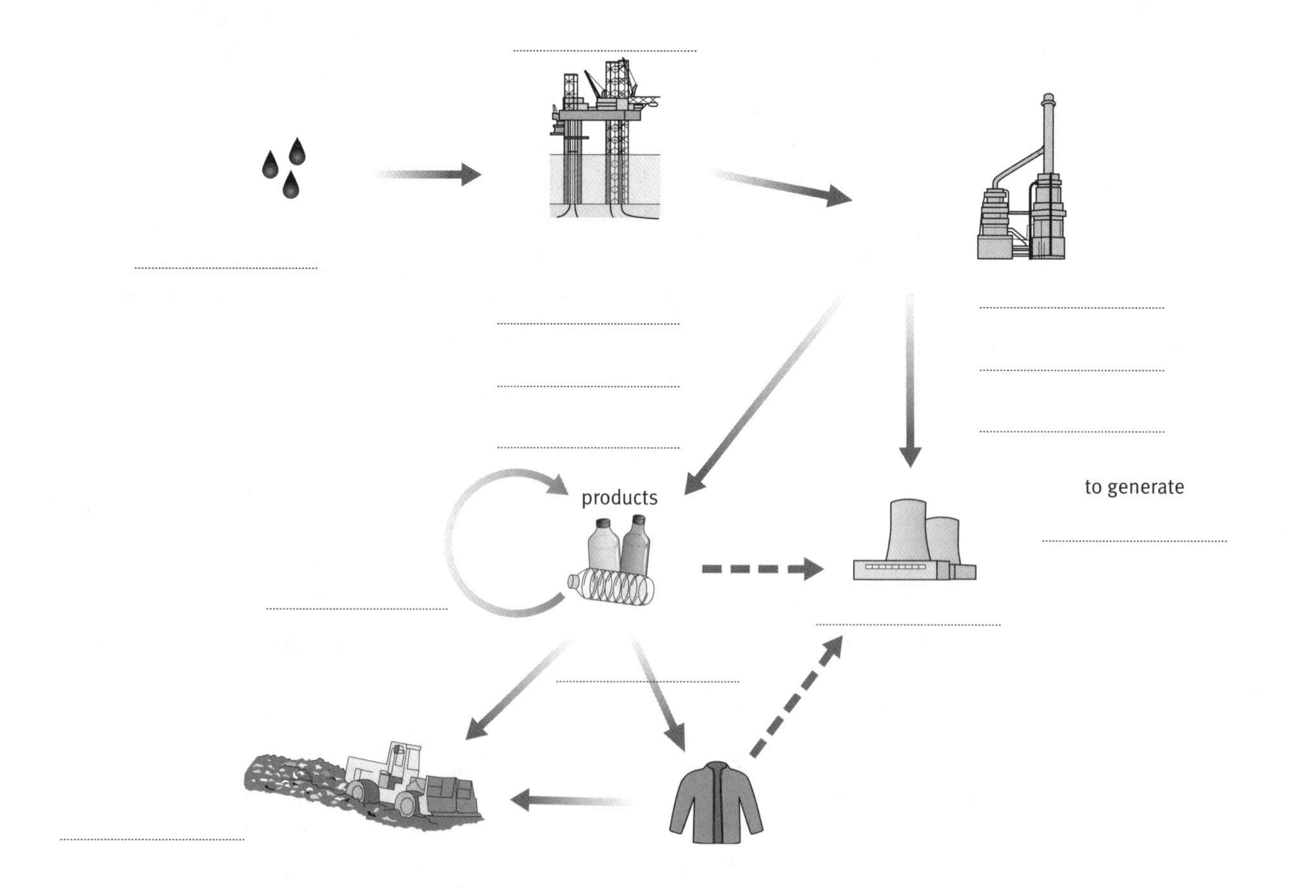

12 Life cycle of a polythene product

This diagram shows the stages in the life of a polythene bottle.

a Write at the bottom of the diagram two ways of disposing of the product other than recycling.

b Write at the top of the diagram the name of the main raw material used to make polythene.

c Draw labelled arrows on the diagram to show how **reusing** the product could give it a different life cycle.

d Identify stages in the life cycle that need an input of energy. Write them below. One has been done for you. Complete the other two.

Stage: *Transport of raw material*
Energy needed: *Fuel for the lorry*

Stage: ______________________

Energy needed: ______________________

Stage: ______________________

Energy needed: ______________________

e Give two environmental problems caused by making polythene products:

- ....
- ....

f Give two benefits of using polythene instead of other materials to make things:

- ....
- ....

g Give two environmental problems caused by disposing of polythene products:

- ....
....
- ....
....

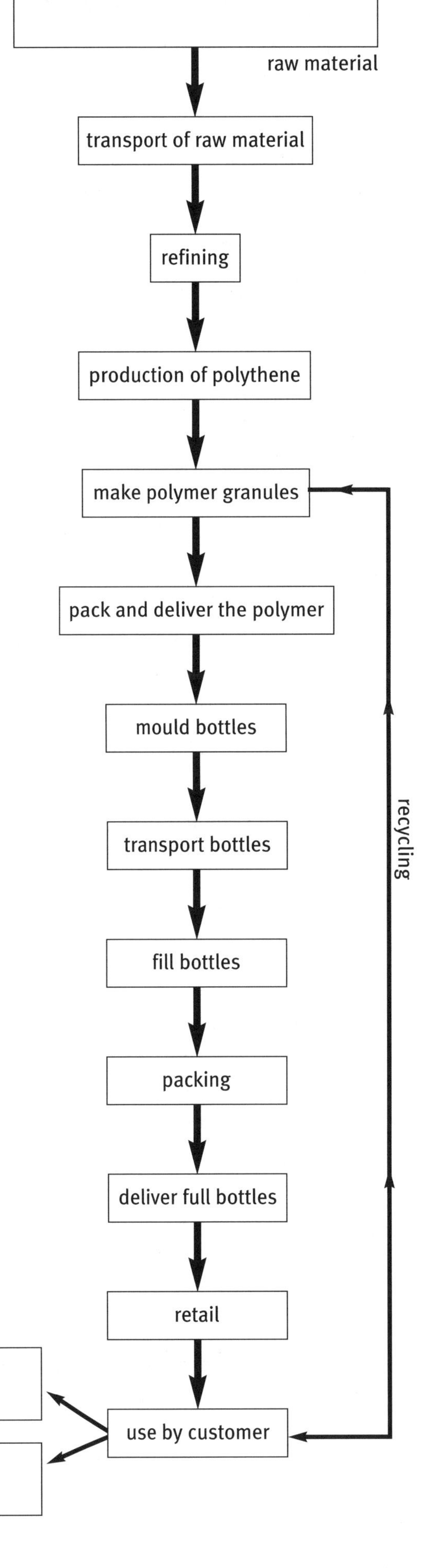

13 Impact on the environment

A life cycle assessment (LCA) measures how a product affects our environment during its whole life. This means from when it is first made to when it is got rid of. All products have some effect (**impact**) on the environment. People have to decide how much impact is acceptable.

The graph compares the impact of three materials that can be used for making water pipes. The impact of PVC is given a rating of 1. The other two materials are compared with PVC.

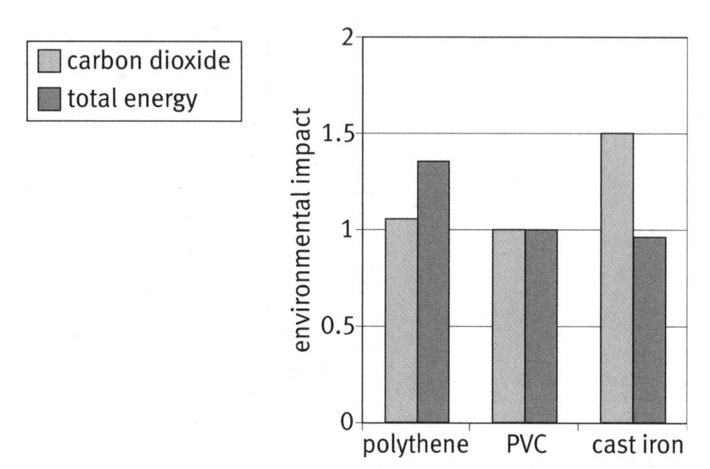

These results come from the Swiss government's lab for materials testing.

a Suggest two examples of when carbon dioxide is made during the life cycle of a plastic water pipe.

➭ ....

➭ ....

b Explain how carbon dioxide can have an impact on the environment.

....

....

....

....

c Write down three things that the graph tells you about the environmental impact of the three materials.

➭ ....

➭ ....

➭ ....

d You are trying to decide which material to use for making water pipes. List two other types of information that you would need before you could do this.

➭ ....

➭ ....

14 Getting rid of waste

The graph shows different ways in which waste was treated in European countries during the 1990s.

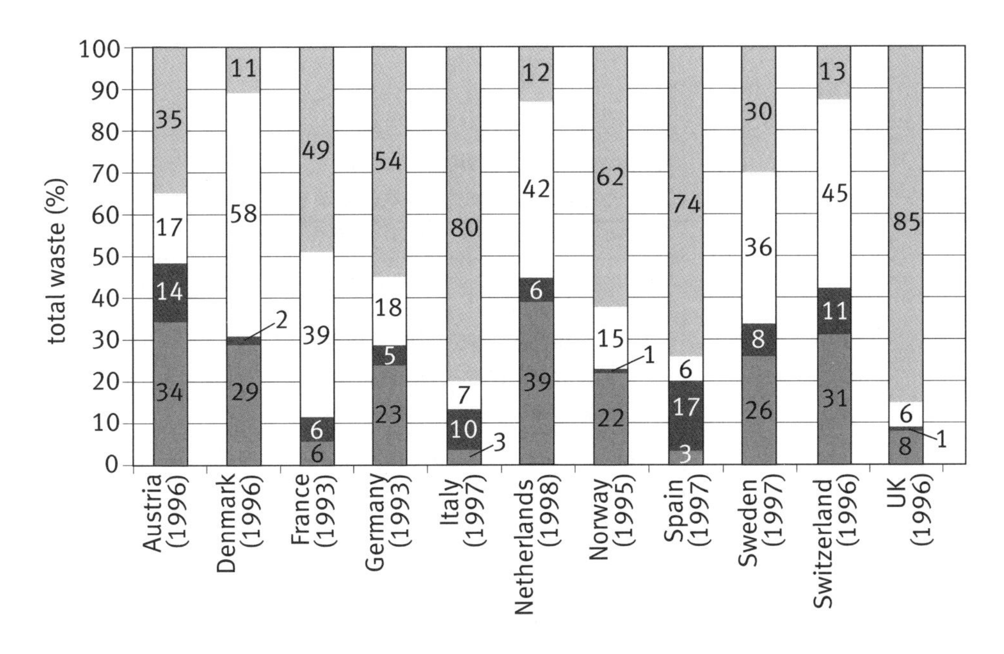

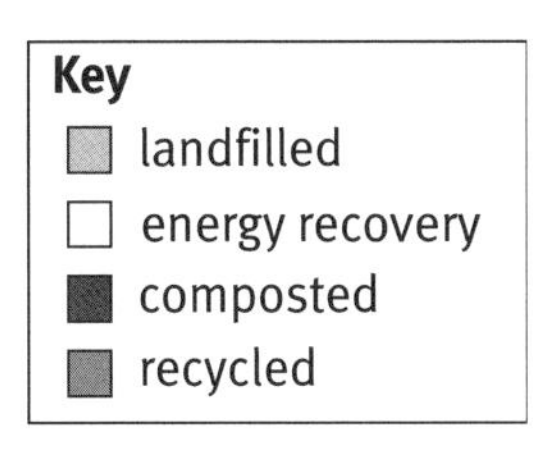

a What was the main way of getting rid of waste in most European countries in the 1990s?

....

b What is the difference between getting rid of waste in the UK and the Netherlands?

....

....

Suggest one reason for this difference.

....

....

c Which country do you think caused least damage to the environment when it got rid of its waste? Explain your choice.

....

....

d Suggest three ways to reduce the amount of waste which countries have to get rid of.

- ....
- ....
- ....

15 Sustainable development

Many people are worried about changes in our environment. They are trying to encourage support for sustainable development.

a Draw lines to match each term with its correct meaning.

sustainable	Changes in our world. For example, changes to how we grow our food, make goods, and organize society.
development	Changes that: ➭ meet everyone's needs ➭ protect the environment ➭ conserve natural resources
sustainable development	Can be done without harming people or the environment.

b Statements **A** to **L** are all about polymers.

A Making synthetic fibres from starch extracted from wheat	**B** Increased use of composite materials with different materials bonded together	**C** Protecting the goods we buy in shops with plastic wrapping film	**D** Making manufacturers responsible for the disposal of the packaging of goods they supply
E Burning waste, including plastic waste, to generate electricity	**F** Dumping unsorted household waste in landfill sites	**G** Using polymer foam to insulate the walls of buildings	**H** Increasing the use of biodegradable plastics
I Making a rule that products can only be made of polymers if the polymers can be recycled easily	**J** Replacing plastic supermarket bags with paper bags	**K** Using sheets of polymer instead of glass for windows in buildings	**L** Making it mandatory that all soft-drinks bottles are designed to be collected and reused

c Which statements are about using polymers in ways that are not sustainable? (Write the letters.)

....

d Which statements are actions we could take to make our use of polymers more sustainable?

....

e Which statements do you need more information about to decide whether or not they improve sustainability?

....

16 Using a scientific approach

- Shade in one colour the speech bubbles with statements that could be checked or investigated by a scientific approach.
- Shade in another colour the speech bubbles with statements that science alone cannot deal with.
- Colour in the key.

Key
☐ A scientific approach can investigate this
☐ Science alone cannot deal with this

Cars made of plastic are lighter and use less fuel

Manufacturers should have to pay for the cost of recycling the goods they make and sell

Clothes made of synthetic fibres are warmer and more weatherproof than clothes made of natural fibres

Crude oil should be saved for making chemicals and plastics – not burnt as a fuel

Recycled plastics are not as strong as freshly made plastics

Clothes made of natural fibres are more fashionable than clothes made of synthetic fibres

17 Costs and benefits

The statements in this table are about applications of science that affect people.

Complete the table to show the groups affected and the main benefits and costs of each application.

Statements	Possible benefits to people and the environment. Who wins?	Possible harm to people and the environment. Who loses?
There is a large refinery and chemical plant near the coast close to Southampton.		
Supermarkets supply fruit in plastic containers covered in plastic film.		
There is a large waste incinerator in Edmonton, North London, where household waste is burnt.		

18 Ways of increasing sustainability

This table has a list of actions that can contribute to sustainability. Complete the table.

Action to increase sustainability	Could it be done? Give a reason	Should it be done? Give a reason	What action by governments, public services, industry, or individuals would make this more likely to happen?
Increase the use of renewable resources such as wood, wool, or seashells and cut down on mineral resources.			
Increase the use of recycled plastics when making products.			
Design products so that they last longer and do not have to be replaced so quickly.			
Design products so that they are easier to reuse or recycle at the end of their lives.			
Design products so that they are cheaper to run and easier to maintain.			
Increase recycling rates for household waste from 25% to 50%.			

1 Food chains

a The words in the boxes describe stages in the food chain for bread from 'field to plate'. Number the boxes in order and draw lines to show the stages in sequence.

☐ ploughing land and adding manure or fertilizer	☐ harvesting the wheat	☐ milling the grain to make flour
☐ planting wheat seeds	☐ mixing the flour with other ingredients, then letting it rise	☐ baking the dough to make bread
☐ protecting the growing crop from pests and diseases	☐ selling bread	☐ storing bread at home
	☐ eating bread	

b Pick three of the stages and write them into the table below. Complete the table to show why people who buy and eat bread might care about each stage. This might be for health reasons, for environmental reasons, or for social or economic reasons.

Stage of the food chain	**Reasons why a consumer might have concerns about this stage**

2 Intensive and organic farming

Complete this table to compare intensive methods of farming with organic farming.

Aspect of farming	Intensive farming	Organic farming
yield of crops from the land		
methods used to keep the soil fertile		
methods used to prevent crops being damaged by pests and diseases		
effects on the environment		
cost and quality of the products		
possible health benefits or risks		
sustainability		

3 The nitrogen cycle

a Use these words and phrases to label the diagram of the nitrogen cycle.

bacteria in the roots of some plants	nitrogen gas in the air	protein in plants
to ground water	animal wastes, dead organisms	nitrates in the soil
decay by bacteria and fungi	protein in animals	manufacture of fertilizers
reactions in thunderstorms	bacteria in the soil that turn nitrates into nitrogen	

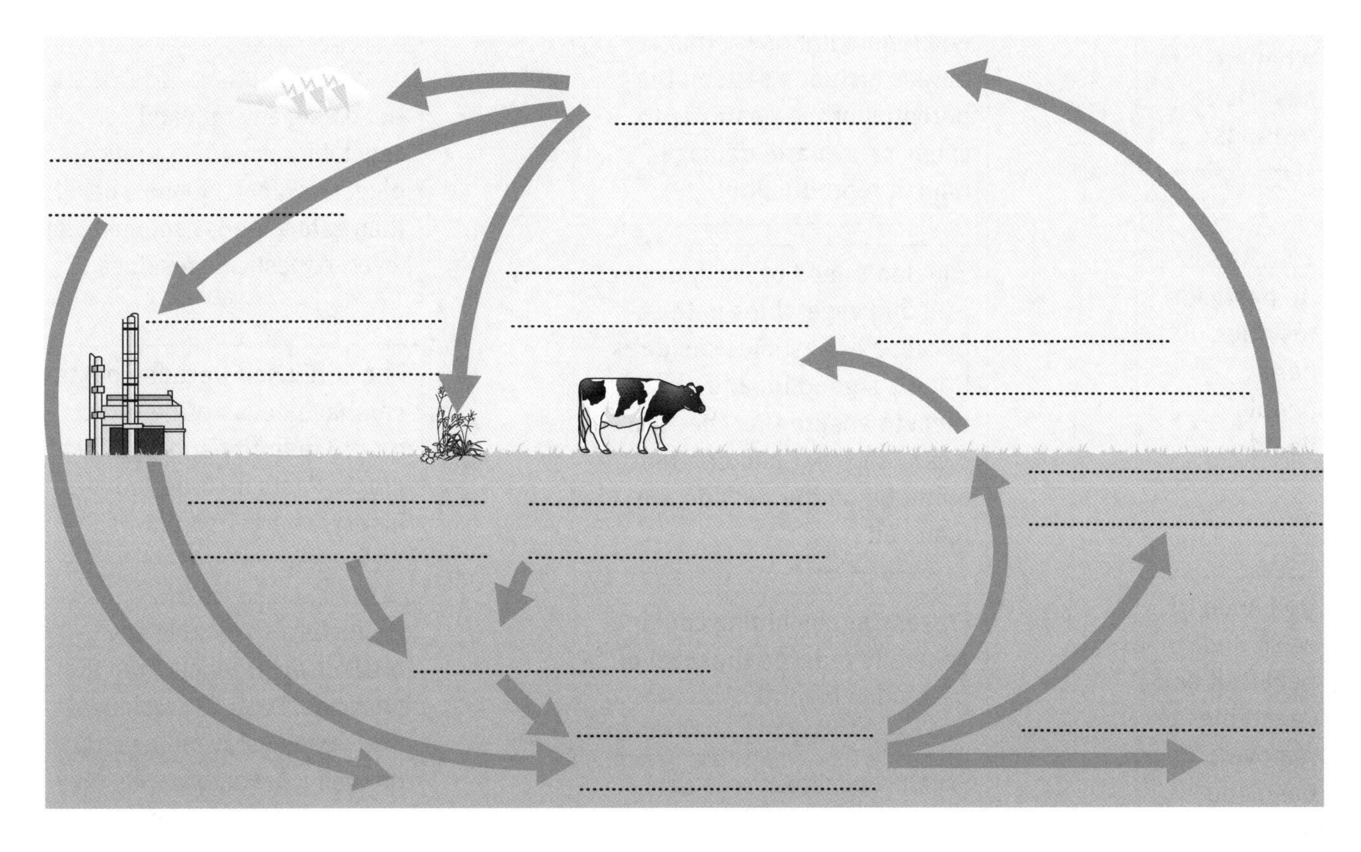

b Explain the following with the help of the labelled diagram to show that human activity can affect the natural nitrogen cycle.

- Harvesting crops tends to reduce the fertility of the soil.

- Use of too much manufactured fertilizer can lead to water pollution.

4 Pesticides and food safety

These questions and answers are based on information provided by the Food Standards Agency on their website. Here the parts of the answers are muddled up.

Draw lines to match each question on the left to the first part of the answer in the middle and the second part of the answer on the right. The first one has been done for you.

What are pesticides?	Any company wanting to get a pesticide approved must submit an application containing information on any potential health and environmental risks. This always includes data on the potential of the pesticide to cause cancer and damage human reproduction.	Rigorous safety assessments are undertaken to make sure that any pesticide residues remaining in the crop will not be harmful to people.
What are pesticide residues?	You don't need to wash or peel fruit and vegetables in the UK because of pesticide residues. But it's a good idea to wash them to ensure that they are clean, and that bacteria that might be on the outside are removed.	Not eating any fruit and vegetables would be a much bigger risk to someone's health than eating foods containing low levels of pesticide residues.
Do pesticide residues harm people's health?	Processing, including cooking, generally reduces the level of pesticides in food.	They are used because pests can have devastating effects on the quantity and quality of crops. Pests include rodents, insects, fungi and plants.
Do I need to wash and peel fruit and vegetables to remove pesticide residues?	Expert committees set safety limits for all approved pesticides, based on scientific evidence. Eating foods containing pesticide residues at levels below the safety limits should not harm people's health.	Residues also include any breakdown products from the pesticide. Pesticide residues may need to stay on the crop to do their job. For example, they may need to be on the surface of a fruit or vegetable to protect it from pests during storage.
Does cooking reduce pesticide residues?	They are chemicals used to kill or control pests that harm our food, health, or environment.	This is because processing can break down the pesticides, or remove the part of the plant that carries the residue.
How are pesticides approved and regulated?	They are the very small amounts of pesticides that can remain in a crop after harvesting or storage and make their way into the food chain.	Washing and peeling may help remove residues of certain pesticides. But some pesticides are systemic, which means they are found within the fruit or vegetable.

5 Food additives

The paragraphs below describe food additives. The headings to the paragraphs are missing. Add the headings using the words in the box.

Antioxidants	Colours	Emulsifiers	Flavourings
Preservatives	Stabilizers	Sweeteners	

➲ ....

These chemicals protect against microorganisms which spoil food and cause food poisoning. Examples are sodium nitrite in ham, calcium propionate in bread and sulfur dioxide in wine and beer.

➲ ....

These chemicals slow down oxidation of oils and fats. Without these chemicals, oily and fatty foods can turn rancid by reacting with the air. Examples are tocopherols in fat for cakes and ascorbyl palmitate in margarine.

➲ ....

These chemicals are used to make foods which have mixtures of fats/oils with water. Examples are lecithin in chocolate and the mono- or diglycerides in ice cream.

➲ ....

These chemicals improve the texture of food. They improve the consistency of food. Examples are pectins in jam, locust bean gum in ice cream and xanthan gum in low-oil dressings.

➲ ....

These chemicals add to the taste of processed foods. A wide range of these chemicals is added to food such as soft drinks, margarine, ice cream and other desserts, soups and sauces.

➲ ....

These chemicals improve the appearance of food after processing. They make manufactured foods look more tasty. Examples are caramel colour in soups, carotenes in cheese and carmoisine in soft drinks.

➲ ....

These chemicals replace sugar in low-calorie foods and provide sweetness without calories.

6 Preservatives

Read these paragraphs about preservatives. Add one or two examples after each paragraph.

- Many foods go off very quickly without preservatives. This means that a lot of food is wasted between the field and the plate, because it goes bad.

  Example(s) ....

  ....

- Traditional preservatives have been used for hundreds of years. Smoking, salt, and vinegar are still used to preserve foods. There are doubts about the safety of some smoked food, because smoke contains a large number of polycyclic hydrocarbons, many of which are known to cause cancer.

  Example(s) ....

  ....

- Preservatives work by killing or stopping the growth of micro-organisms. Micro-organisms cannot live and grow in food that is too acidic or too sweet.

  Example(s) ....

  ....

- Sulfur dioxide is very widely used as a preservative. It has been added to foods for a very long time. More modern preservatives, such as potassium sorbate and sodium benzoate, act like antibiotics. They stop bacteria growing. Most of the preservatives are simple chemicals that are closely related to natural substances.

  Example(s) ....

  ....

- Preservatives work by preventing the formation of micro-organisms, some of which are exceedingly dangerous. Preservatives themselves are generally safe.

  Example(s) ....

  ....

7 Antioxidants

Read the text in the box and then complete the exercises and notes below.

> Oxygen in the air can spoil food. The oxygen does this by reacting with chemicals in food. A common example is the browning of apples or potatoes when cut open in the air. A drop of lemon juice can stop the browning. Lemon juice contains vitamin C (E300), which is a good antioxidant.
>
> Antioxidants prevent a lot of food wastage.
>
> Vegetables contain several antioxidants. Two examples are vitamins C and E.
>
> The most common synthetic antioxidants are butylated hydroxyanisole (BHA; E320) and butylated hydroxytoluene (BHT; E321). These chemicals have been used for many years but not everyone is happy about their use.
>
> The same chemical properties which make BHA and BHT excellent preservatives, may also mean that they have health effects. Some people may have difficulty metabolizing BHA and BHT, which can lead to health and behaviour changes. Some tests in animals indicate that large doses of BHA or BHT can cause cancer. Other studies show that antioxidants can help to protect against cancer by destroying other harmful chemicals.
>
> Experts belonging to an international committee set up by the Food and Agriculture Organization (FAO) and the World Health Organization (WHO) have studied the research information about BHA and BHT. They have set Acceptable Daily Intakes (ADIs) for them at 0–0.5 mg/kg body weight for BHA and 0–0.3 mg/kg body weight for BHT.

a Explain why antioxidants are needed to stop food going off.

....

....

b Give two examples of antioxidants.

- A natural antioxidant: ....
- A synthetic antioxidant: ....

c Why do the FAO and WHO set ADIs for some antioxidants?

....

....

d How can experts decide on the values to set for ADIs?

....

....

8 Emulsifiers

Read the text in the box and then answer the questions below.

Milk is a natural emulsion. It is a complicated mixture of fat droplets in a watery solution.

In cooking, one of the most common emulsifiers is a chemical in eggs called lecithin (E322). It is the lecithin in egg yolks that makes it possible to make mayonnaise. Mayonnaise is an amazing emulsion. It consists of 80% oil mixed up with 20% of an acidic solution in water.

Molecules of emulsifiers have one oil-friendly end and one water-friendly end. They act as a bridge between oil and water. They make it possible for oil and water to mix.

Some emulsifiers used in food production come from natural products. Others are synthetic chemicals with a very similar structure to the natural products.

a What happens if you shake up a vegetable oil and water together without adding an emulsifier?

....

....

b Give two examples of foods that could not exist if they did not contain emulsifiers.

- ....
- ....

c Why do some people prefer emulsifiers such as lecithin?

....

....

---

9 Additives and E numbers

Draw lines to join the start of each sentence with the correct ending.

An additive is given an E number to . . .	. . . avoid eating additives that they do not want to eat.
EU legislation requires food labels to . . .	. . . make sure that food manufacturers using additives do not use too much or chemicals that are illegal.
Looking at food labels allows customers to . . .	. . . give the names or E numbers of additives used.
The Food Standards Agency checks up on additives to . . .	. . . show that it has passed safety tests and been approved for use in the European Union.

10 Natural polymers

Use these words to complete the captions for the two diagrams below.

Then colour the atoms in the models:

carbon – black   oxygen – red   nitrogen – blue   hydrogen – white

amino acids	carbohydrate	carbon	glucose	hydrogen	nitrogen
oxygen	polymer	polymers	proteins	starch	sugar

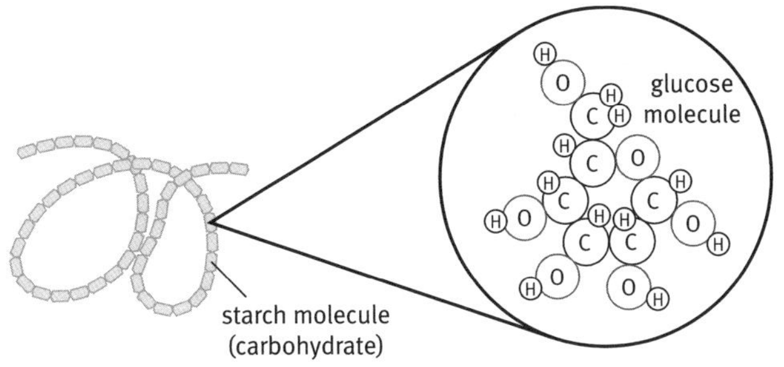

amino acid molecule

protein molecule

Starch is an example of a ………………………… Other carbohydrates are ………………………… and …………………………. Starch is a ………………………… made by joining up ………………………… molecules in a long chain. The elements in carbohydrates are …………………………, …………………………, and oxygen.

Proteins are ………………………… made up of long chains of …………………………. The elements in ………………………… and amino acids are mainly carbon, hydrogen, …………………………, and ………………………….

11 Digestion

a Explain why your body has to digest natural polymers in food such as starch and protein.

……………………………………………………………………………………

……………………………………………………………………………………

b Explain what happens to the chemicals produced by digestion.

……………………………………………………………………………………

……………………………………………………………………………………

12 Proteins in the body

Use these words to complete the labelling of this diagram to show which parts of the body consist of protein.

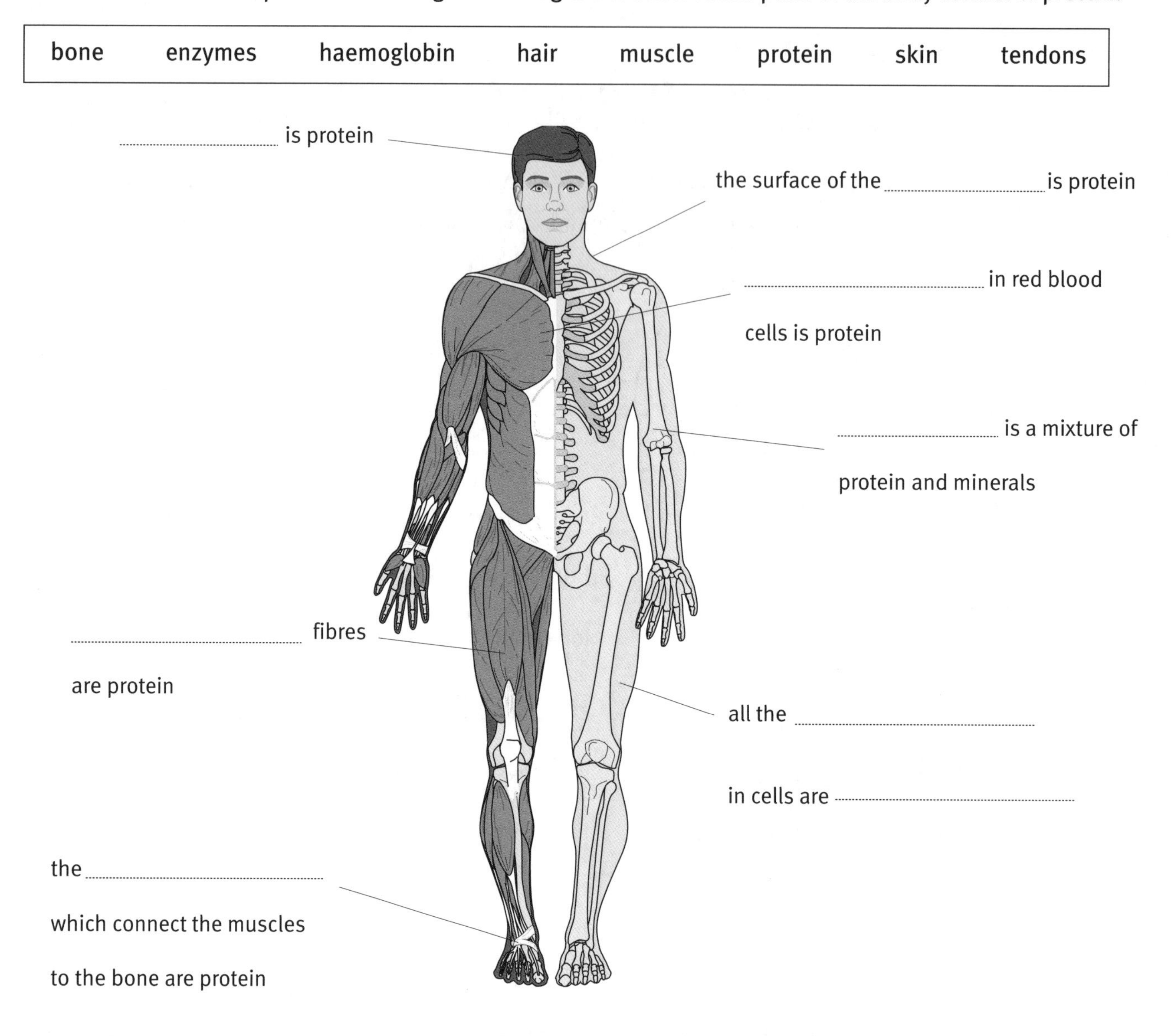
bone enzymes haemoglobin hair muscle protein skin tendons
.... is protein
the surface of the .... is protein
.... in red blood cells is protein
.... is a mixture of protein and minerals
.... fibres are protein
all the .... in cells are ....
the .... which connect the muscles to the bone are protein

13 Getting rid of unwanted amino acids

Complete the labelling of this diagram to describe what happens in your body if you eat more amino acids than you need.

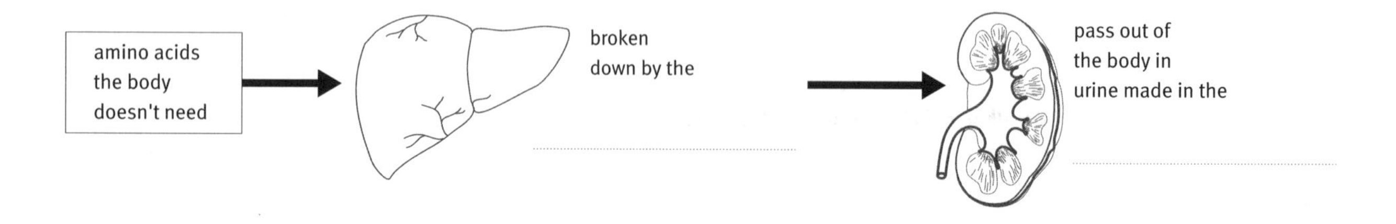
amino acids the body doesn't need
broken down by the ....
pass out of the body in urine made in the ....

14 The chemicals in cola

These two lists of ingredients are for different types of cola. One is ordinary cola. The other is 'diet' cola.

a Complete the heading on each label to show which is which.

.... cola	.... cola
carbonated water, colour (E150b), flavouring (including caffeine), phosphoric acid, sweeteners (aspartame, acesulfame K), acidity regulator (sodium citrate), preservative (sodium benzoate), citiric acid	carbonated water, sugar, colour (E150d), phosphoric acid, flavourings (including caffeine)

b Explain how you can tell which label is which.

....

....

....

....

c Suggest a reason why some people prefer to drink ordinary cola.

....

....

d Suggest a reason why some people prefer to drink diet cola.

....

....

15 Blood sugar levels

There are two lines on this graph, which shows the level of sugar in the blood after two meals.

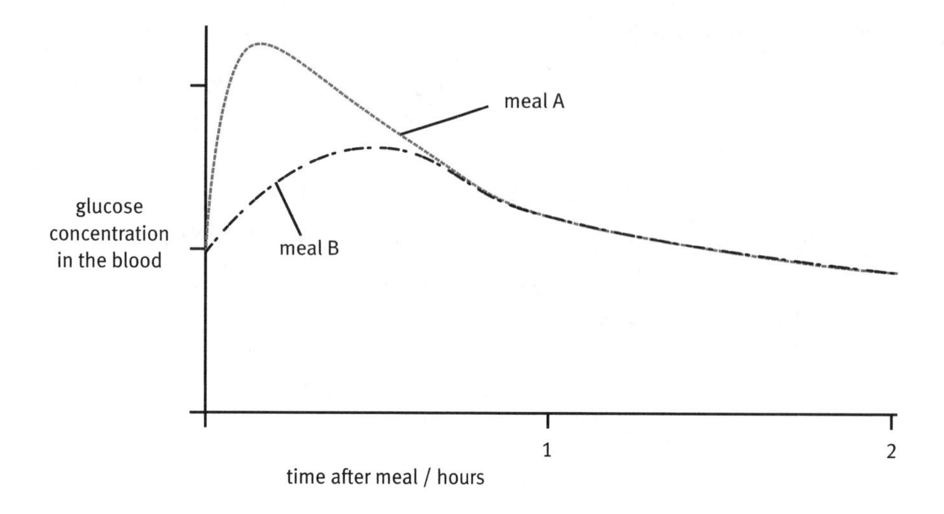

a What do the lines on the graph tell you about what happens to the level of sugar in the blood after a meal?

....

....

b Why does the level of sugar in the blood change after a meal in the way you have described in **a**?

....

....

....

c Which line shows what happens after a meal with a lot of sugary food? ....

d Look at the other line. Suggest examples of food that might have been in that meal.

....

....

16 Diabetes

The grid contains three types of statement related to diabetes.

Key

☐ Statements mainly about Type 1 diabetes

☐ Statements mainly about Type 2 diabetes

☐ Statements which are true for both kinds of diabetes

➲ Use different colours to complete the key above.

➲ Then use the same colours to identify the different types of statement in the grid.

Glucose comes from sugar and other sweet foods.	This type of diabetes usually appears before the age of 40.	The body can still make some insulin, but not enough.	The level of glucose (sugar) in the blood is too high if people who have diabetes are not treated.
With some people this can be treated by changes to lifestyle, such as a healthier diet, weight loss, and more exercise.	The aim of treatment is to keep blood glucose levels as near to normal as possible.	Tablets and/or insulin may be needed for normal blood glucose levels.	The body is unable to produce any insulin.
Insulin is a hormone produced by the pancreas.	The insulin that is produced does not work properly.	Glucose comes from the digestion of starchy foods.	The condition develops quickly and the symptoms are obvious.
This disease mainly affects older people.	The symptoms develop slowly and may not be noticed for quite a while.	Diabetes in some people can only be treated by insulin injections.	Most people affected in this way are overweight.

17 Food risks and regulations

Complete this table to illustrate examples of risk and regulation along the food chain from field to plate. Give one example in each empty box.

Stage in the food chain	**Risks to health from chemicals or other sources**	**Ways that government and regulators try to protect the public**
Natural chemicals in growing crops		
Cultivating and harvesting crops		
Transporting and storing crops		
Preserving and processing food		
Cooking and serving food		
Allergies that can affect some people		
Eating an unhealthy diet		

18 The Food Standards Agency

The following text is based on information from the Food Standards Agency. Read the text and then answer the questions.

> Acrylamide is a chemical that was found in unexpectedly large amounts in starchy foods that had been cooked at high temperatures. These included crisps, chips, bread, and crispbreads. Acrylamide causes cancer in animals and so may also harm people's health. The Food Standards Agency has carried out its own research, which confirmed the original findings made in Sweden in 2002.
>
> The Agency's research included tests on pre-cooked, processed, and packaged foods, plus chips that were prepared from potatoes and cooked by the researchers. High levels were found in the home-cooked foods and in the processed foods.
>
> The research also investigated how acrylamide is formed and how acrylamide levels can be reduced. The main aim is to find out how to minimize the amount of acrylamide present in food.
>
> There is no general limit set for acrylamide in food because levels of this sort of chemical should be kept as low as is reasonably practical. There is a legal limit set for acrylamide from plastics used in contact with food, such as packaging.

a Why did the Food Standards Agency repeat the research that had already been done in Sweden?

....

....

b What else would you want to ask the researchers to investigate if you were planning the next stage of research by the Food Standards Agency?

....

....

....

c What risks might there be if people tried to avoid forming acrylamide by cooking food at a lower temperature?

....

....

....

d Suggest what you could do if you wanted to adopt a precautionary approach to the problem of acrylamide in food while waiting for the experts to give final advice.

....

....

....

1 The elements in order

Complete the sentences and diagram.

Scientists look for patterns in data. When they arrange the known elements in order of relative ...., they find that there is a repeating pattern. These patterns are shown clearly when the elements are arranged in a .... Each row in the table is a called period, with metals on the .... and .... on the right.

A column of elements with similar properties

G

P __ __ __ __ D

Vertical columns of elements in the table are .... of elements with similar properties.

Modern versions of the periodic table put the elements in order of .... number, also known as the atomic number.

2 The periodic table

a Colour the key, then use the colours to show these parts on the periodic table below.

☐ group 1 (alkali metals) ☐ group 7 (halogens) ☐ non-metals ☐ transition metals

b Over three-quarters of the elements are metals. Lightly colour in the metals.

Key: relative atomic mass; symbol; name; proton number (e.g. 1, H, hydrogen, 1)

group number / period number

period	1	2											3	4	5	6	7	8
1	1 H hydrogen 1																	4 He helium 2
2	7 Li lithium 3	9 Be beryllium 4											11 B boron 5	12 C carbon 6	14 N nitrogen 7	16 O oxygen 8	19 F fluorine 9	20 Ne neon 10
3	23 Na sodium 11	24 Mg magnesium 12											27 Al aluminium 13	28 Si silicon 14	31 P phosphorus 15	32 S sulfur 16	35.5 Cl chlorine 17	40 Ar argon 18
4	39 K potassium 19	40 Ca calcium 20	45 Sc scandium 21	48 Ti titanium 22	51 V vanadium 23	52 Cr chromium 24	55 Mn manganese 25	56 Fe iron 26	59 Co cobalt 27	59 Ni nickel 28	63.5 Cu copper 29	65 Zn zinc 30	70 Ga gallium 31	73 Ge germanium 32	75 As arsenic 33	79 Se selenium 34	80 Br bromine 35	84 Kr krypton 36
5	86 Rb rubidium 37	88 Sr strontium 38	89 Y yttrium 39	91 Zr zirconium 40	93 Nb niobium 41	96 Mo molybdenum 42	97 Tc technetium 43	101 Ru ruthenium 44	103 Rh rhodium 45	106 Pd palladium 46	108 Ag silver 47	112 Cd cadmium 48	115 In indium 49	119 Sn tin 50	122 Sb antimony 51	126 Te tellurium 52	127 I iodine 53	131 Xe xenon 54
6	133 Cs caesium 55	137 Ba barium 56	139 La lanthanum 57	179 Hf hafnium 72	181 Ta tantalum 73	184 W tungsten 74	186 Re rhenium 75	190 Os osmium 76	192 Ir iridium 77	195 Pt platinum 78	197 Au gold 79	201 Hg mercury 80	204 Tl thallium 81	207 Pb lead 82	209 Bi bismuth 83	210 Po polonium 84	210 At astatine 85	222 Rn radon 86
7	223 Fr francium 87	226 Ra radium 88	227 Ac actinium 89	104	105	106	107	108	109	110	111	112						

3 Periodic patterns

This is a plot of melting point against proton number for some elements.

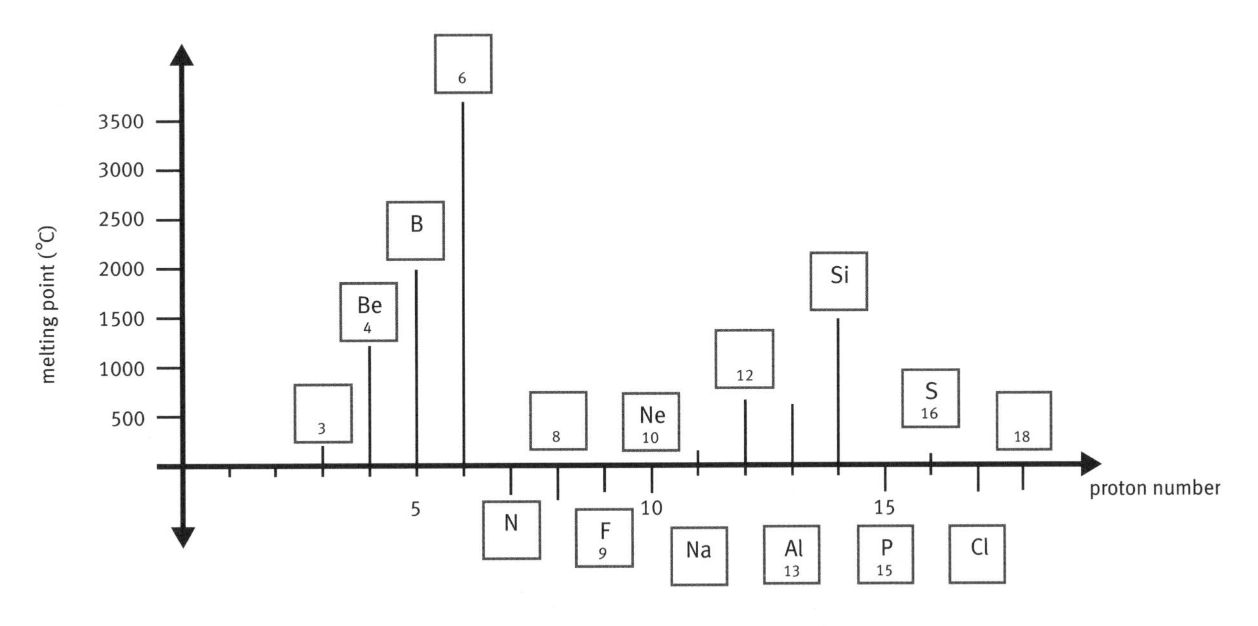

a Use the periodic table on page 48 to fill in the five missing proton numbers and five missing symbols.

b Which two elements are at the peaks? ....

To which group do these two elements belong? ....

c Describe the pattern of melting points across the two periods.

....

This is a plot of melting point against proton number for a group of elements.

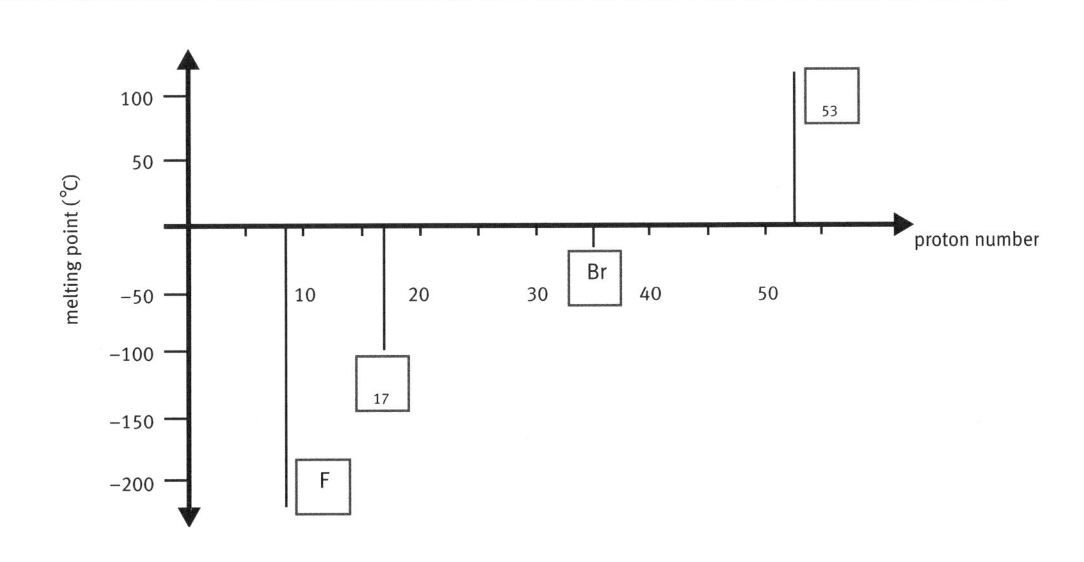

d Use the periodic table on page 48 to fill in the two missing proton numbers and two missing symbols.

e To which group do these two elements belong? ....

f Describe the pattern of melting points down the group.

....

4 The alkali metals

a Complete the information about group 1 elements. The alkali metals:

- are .... – you can cut them with a knife
- are .... – but only when freshly cut
- quickly .... in moist air – they react with water and oxygen
- have low .... – some float on water
- react with water to form .... and an .... solution of the metal ....

e.g. sodium + water → sodium hydroxide + hydrogen

lithium + ... → .... + ...

We call group 1 the **alkali** metals because they produce alkaline solutions when they react with water. They also react vigorously with chlorine. The products are **colourless** crystalline salts called metal chlorides.

e.g. sodium + chlorine → sodium chloride

potassium + ... → ....

b Explain the precautions you must take when using group 1 metals and alkalis.

....

....

c From the information in the table:

Alkali metal	Proton number	Melting point (°C)	Boiling point (°C)	Density (g/cm^3)
lithium	3	181	1331	0.54
sodium	11	98	890	0.97
potassium	19	63	766	0.86

- How does melting point vary with proton number in group 1?

....

- How does boiling point vary with proton number in group 1?

....

- What further information do you need to decide whether or not there is a pattern to the densities of group 1 metals?

....

d In the table below, describe the reactions of alkali metals with water (right-hand column). Choose words from this list.

violent reaction	makes sparks	gas catches fire	
moves around on water	fizzes gently	makes sparks	melts
metal thrown off surface	floats	gives off hydrogen	

Write these words into the left-hand column to show the trend in reactivity down the group:

most reactive	least reactive

Reactivity	Name of metal	Description of reaction with water
	lithium	
	sodium	
	potassium	

5 Names and formulae

Complete the table, with the help of the periodic table on page 48.

Chemical	Symbols of the element(s) in it	Formula of the chemical
hydrogen	H	H_2
water	........, ...	
lithium fluoride	Li, ...	
sodium chloride	Na, ...	
potassium bromide	K, ...	
lithium hydroxide	Li, ..., ...	

6 Balanced equations

a Use these words and symbols to complete the sentences.

g	aq	s	arrow	l	balanced	atoms

In a symbol equation, the number of .... of each element on each side of the .... in an equation must be equal. We call this a .... equation.

We can also show the reactants and products as solid (......), liquid (......), gas (......), or aqueous solution (......).

b Complete the equations to describe the reaction between potassium and water.

Step 1: Describe the reaction in words:

potassium + water → .... + ...

Step 2: Write down the formulae for the reactants and products:

.... + ... → .... + ...

Step 3: Balance the equation:

.... + ... → .... + ...

Step 4: Add state symbols:

.... + ... → .... + ...

c Complete the equations to describe the reaction between lithium and chlorine.

Step 1: Describe the reaction in words:

lithium + chlorine → ....

Step 2: Write down the formulae for the reactants and products:

.... + ... → ....

Step 3: Balance the equation:

.... + ... → ....

Step 4: Add state symbols:

.... + ... → ....

7 The halogens

a Complete these sentences by drawing a (ring) around the correct **bold** words.

- The halogens are **metals / non-metals** and their vapours are **coloured / colourless.**
- Halogens can **colour / bleach** vegetable dyes and kill bacteria.
- The halogens are **toxic / non-toxic** to humans.
- Halogen molecules are each made of **one / two** atoms; they are **monatomic / diatomic.**
- Halogens react with **metal / non-metal** elements to form crystalline compounds that are salts.
- The halides of alkali metals are **coloured / colourless** salts such as **potassium / iron bromide.**
- Compounds of group 1 elements have the formula **MX / MX_2** (where M = metal and X = halide).

b Complete the table. Choose the missing temperatures in the second and third rows from these values:

−34 °C	58 °C	114 °C

Name	chlorine	bromine	iodine
State at room temperature			
Melting point (°C)	−101	−7	
Boiling point (°C)			184

c Describe the reactions of these halogens with hot iron. Draw an arrow to show decreasing reactivity down the group.

Reactivity	**Name of metal**	**Description of the reaction with hot iron**
	chlorine	
	bromine	
	iodine	

8 Flame colours and spectra

a Colour these flames to show the results of a flame test with each of the named salts.

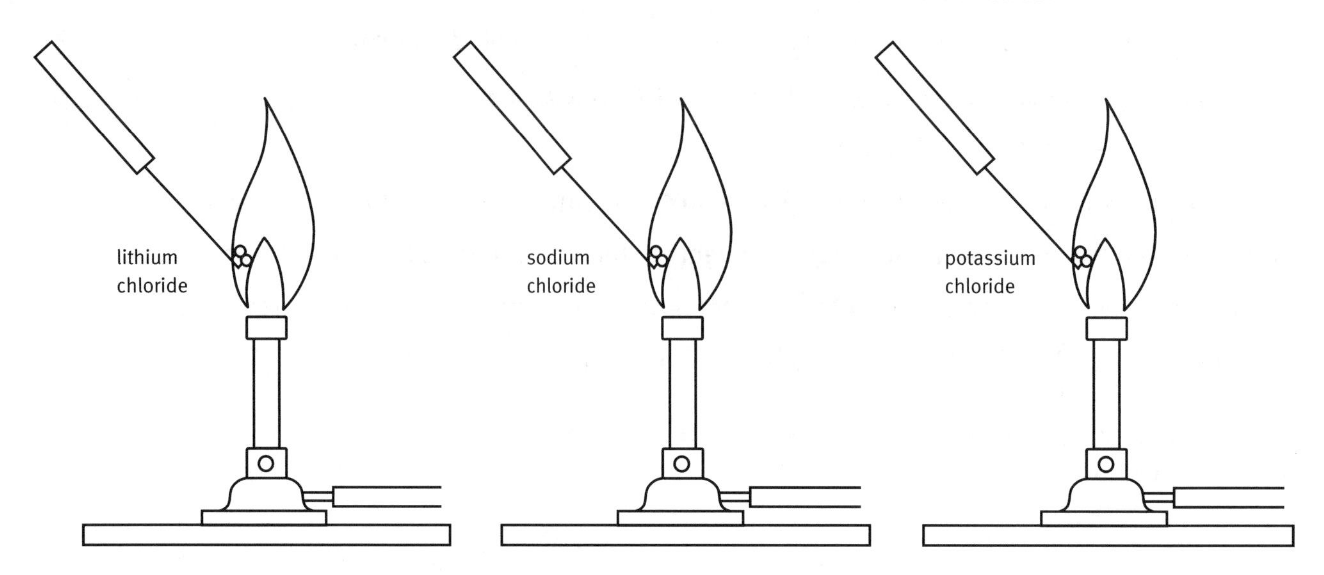

b Colour the lines in this spectrum of helium gas.

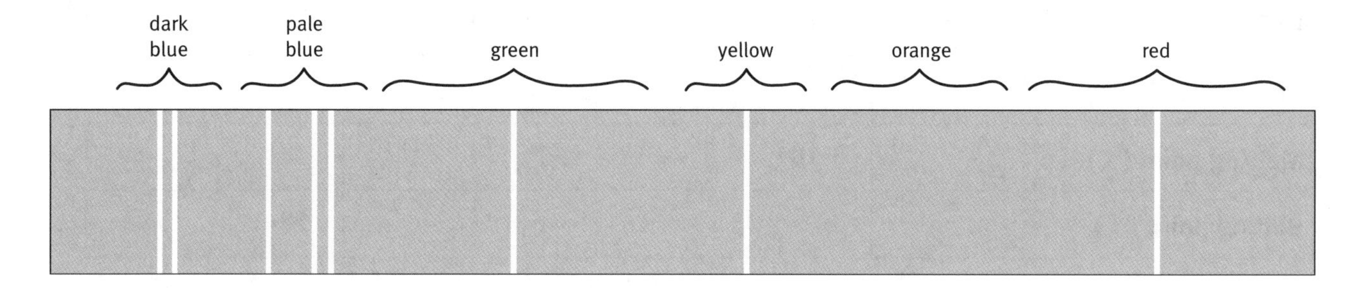

c How does the spectrum of helium differ from the spectrum of white light from the Sun?

....

....

d Why is it possible to use spectra to identify elements during chemical analysis?

....

....

e How was it possible to discover helium on the Sun before it was discovered on Earth?

....

....

....

9 Atomic structure

a All atoms are made of the same basic parts – protons, neutrons, and electrons.

Label the diagram of an atom.

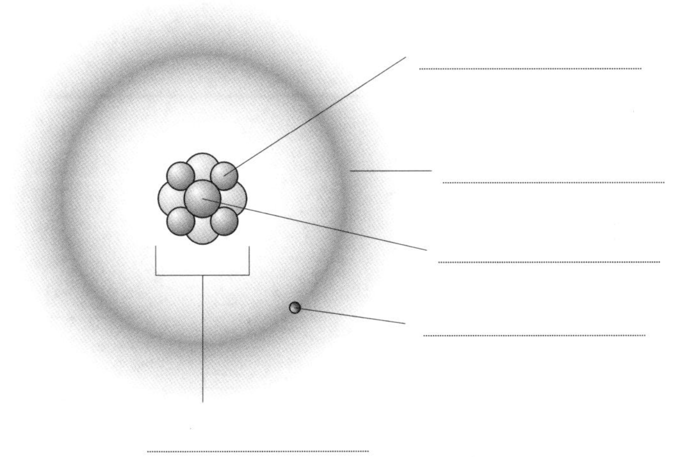

b Complete the table.

Part of atom	Relative mass	Charge	Position
proton		+1	in a very small central nucleus
	1		
electron			in shells around the atom's nucleus

c Complete the sentences using these words.

electrons	proton	protons	charge	atomic

- All atoms of the same element have the same number of ....
- The number of protons is called the .... number or .... number.
- The number of protons is equal to the number of .... This means that an atom has no overall ....

10 Electrons in atoms

a Complete these sentences by filling in the blanks with words or numbers.

The electrons in an atom are arranged in a series of .... around the nucleus. These shells are also called .... levels. In an atom the .... shell fills first, then the next shell, and so on.

There is room for

- up to .... electrons in shell one
- up to .... electrons in shell two
- up to .... electrons in shell three

Shells fill from .... to .... across the .... of the periodic table.

- Shell one fills up first from .... to helium.
- The second shell fills next from lithium to ....
- Eight .... go into the third shell from sodium to argon.
- Then the fourth shell starts to fill from potassium.

electron
nucleus
Na
electron shells

The .... number of sodium is 11. So there are .... electrons in a sodium atom.

The diagram above shows the arrangement of .... in a sodium atom. This electron arrangement can also be written: ....

b Show the arrangements of electrons in these atoms. Sodium has been done for you. You will find the proton numbers in the table on page 48.

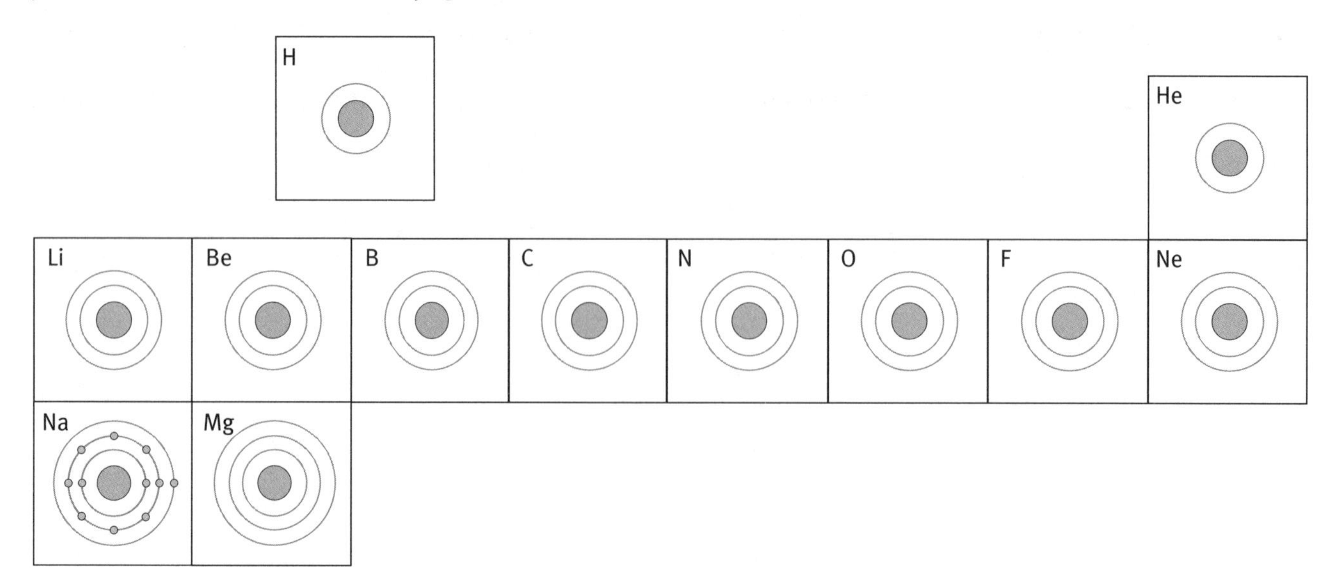

c Complete the lists of the arrangement of electrons (electronic configurations) for:

Alkali metals (group 1 in the periodic table)		**Halogens (group 7 in the periodic table)**	
lithium	2.1	fluorine	....
sodium	....	chlorine	....
potassium	....	bromine	2.8.18.7

d Since it is the electrons in the outer 'shell' that affect chemical reactions, the number of outer-shell electrons determines the chemical properties of an element.

Complete these sentences.

- Group 1 metals have similar properties because they have .... electron in their outer shell.
- Halogens have .... properties because they have .... electrons in their outer shell.

e Give examples or more details to illustrate or justify the general statements made in these two paragraphs.

When atoms react it is the electrons in the outer shell which get involved as chemical bonds break and new chemicals form. Elements have similar properties if they have the same number and arrangements of electrons in the outer shells of their atoms.

....

....

....

....

The alkali metals are not all the same because their atoms differ in the number of inner full shells. A sodium atom has two inner filled shells, so it is larger than a lithium atom and its outer electron is further away from the nucleus. As a result, the two metals have similar but not identical physical and chemical properties.

....

....

....

....

11 Salts and their properties

Complete this table to give examples to illustrate the properties of salts.

Description of salts	Example
Salts are compounds of metals with one or more non-metals.	
Salts are crystalline when solid.	
Salts have much higher melting and boiling points than compounds made up of small molecules.	
Some salts are soluble in water.	
Some salts are insoluble in water.	
Salts do not conduct electricity when solid but they do when molten (liquid). There are changes at the electrodes when an electric current flows through a molten salt.	
Solutions of salts in water conduct electricity. There are changes at the electrodes when an electric current flows through a solution of a salt.	

12 Ionic theory

a The diagram shows a model of the structure of sodium chloride.

Colour the diagram to show:

- sodium ions **red**
- chloride ions **green**

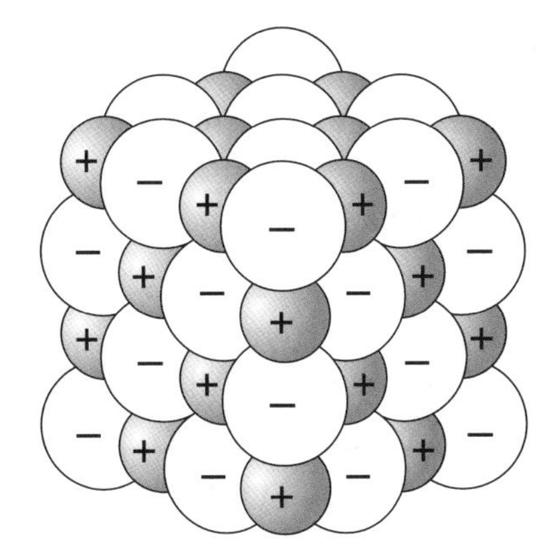

How does this model of the structure of sodium chloride explain the shape of sodium chloride crystals?

....

....

....

b Complete this table.

Property of salts	Ionic theory explanation
All the crystals of each solid ionic compound are the same shape. Whatever the size of the crystal, the angles between the faces of the crystal are always the same.	
	The giant ionic structure is held together by the strong attraction between the positive and negative ions. It takes a lot of energy to break down the regular arrangement of ions.
	In a molten ionic compound the positive and negative ions can move around independently.
The solution of an ionic compound in water is a good conductor of electricity.	

13 Atoms into ions

a Complete the sentences next to these diagrams.

Atoms of metals on the left-hand side of the periodic table turn into ions by losing electrons.

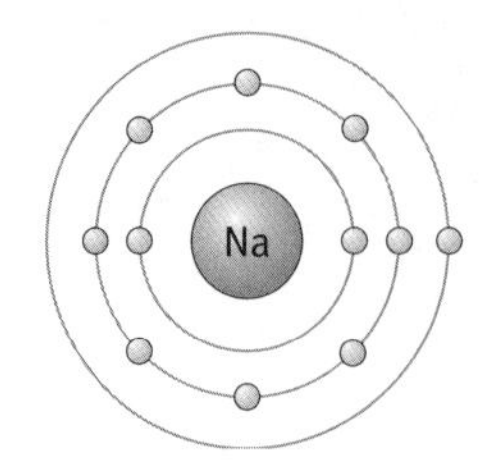

When it turns into an ion the sodium atom .... 1 electron (negative charge)

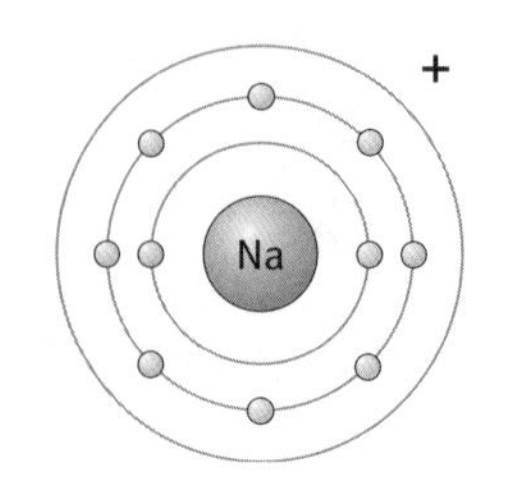

so the sodium ion has a .... charge

Atoms of non-metals to the right of the periodic table turn into ions by gaining electrons.

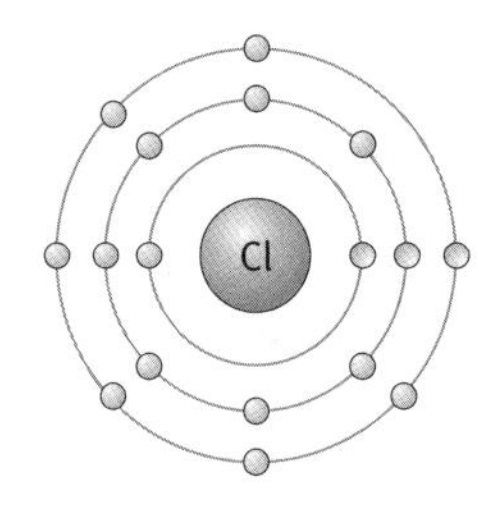

When it turns into an ion the chlorine atom .... 1 electron (negative charge)

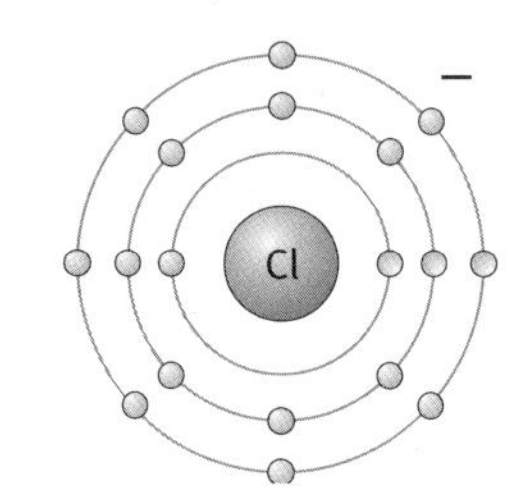

so the chloride ion has a .... charge

b Complete these diagrams to show the number and arrangement of electrons in each atom and the ions they form. (Your answers to question 10b will help you.)

Atom		**Ion**	
Symbol: Li	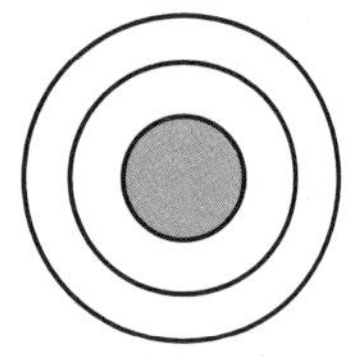	Symbol: Li^+	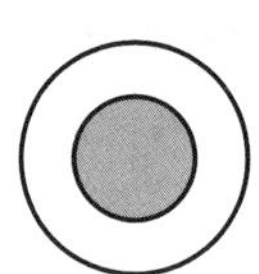
Symbol: Mg	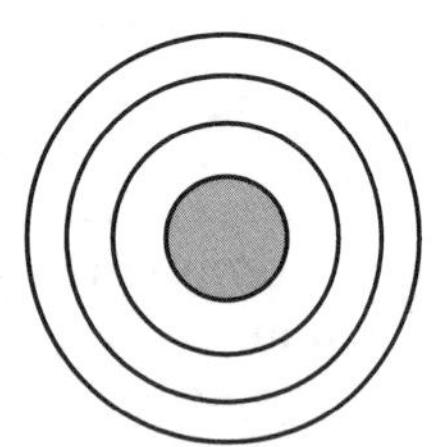	Symbol:	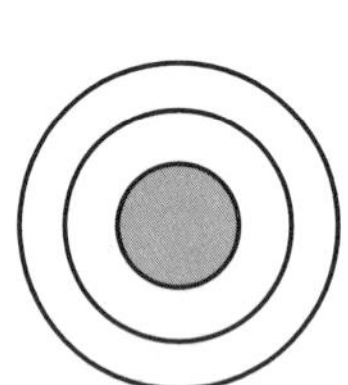
Symbol: F	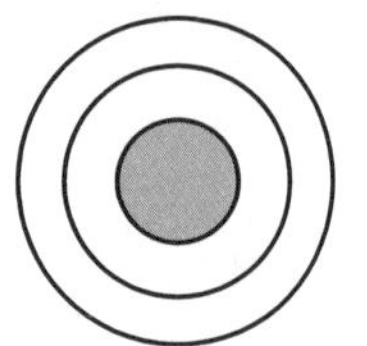	Symbol:	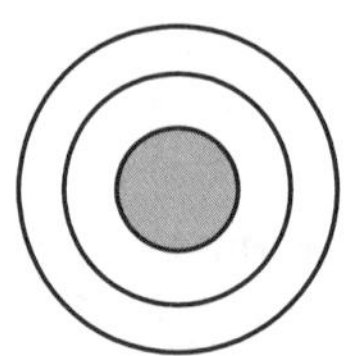
Symbol: O	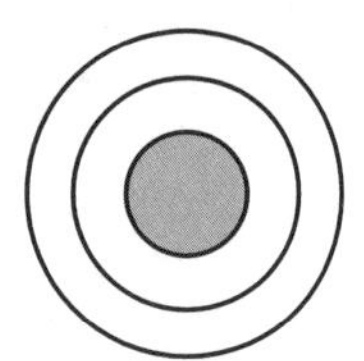	Symbol:	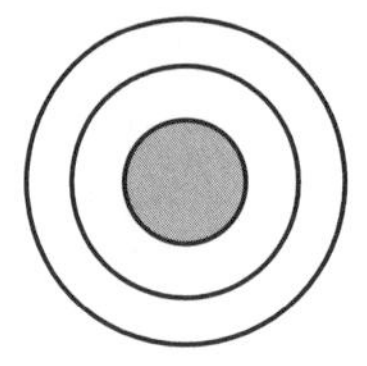

14 Formulae of ionic compounds

This table shows formulae of simple ions.

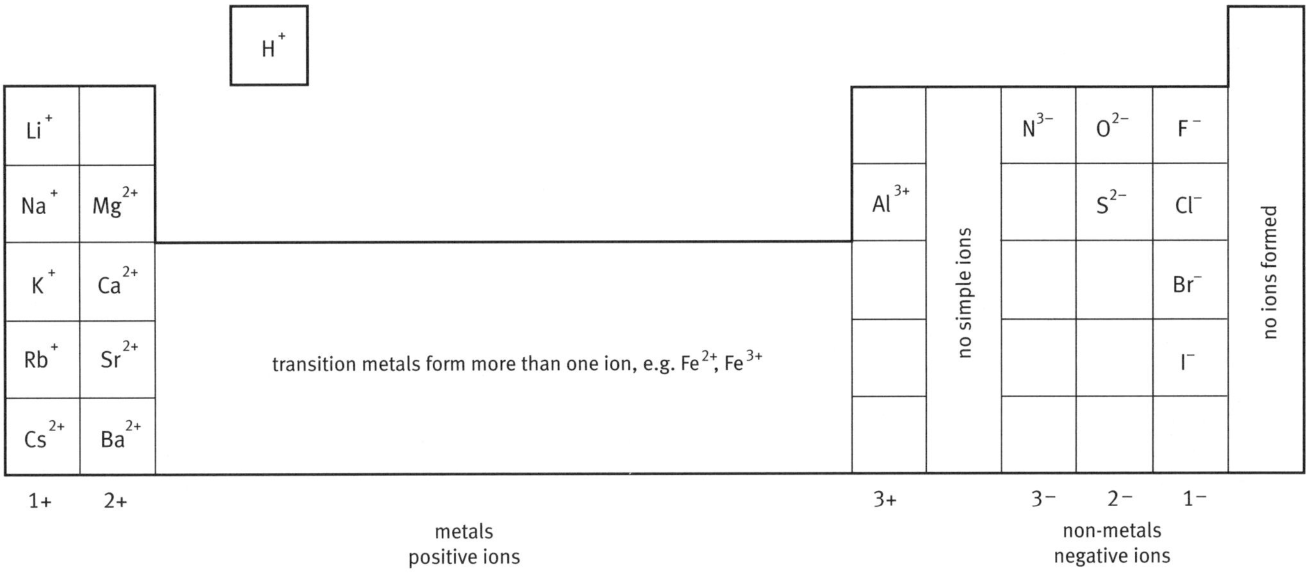

		H^{+}						
Li^{+}				no simple ions	N^{3-}	O^{2-}	F^{-}	no ions formed
Na^{+}	Mg^{2+}		Al^{3+}			S^{2-}	Cl^{-}	
K^{+}	Ca^{2+}						Br^{-}	
Rb^{+}	Sr^{2+}	transition metals form more than one ion, e.g. Fe^{2+}, Fe^{3+}					I^{-}	
Cs^{2+}	Ba^{2+}							
1+	2+		3+		3−	2−	1−	

metals
positive ions

non-metals
negative ions

a Complete these general statements about ions.

- The metals in group 1 and 2 form ....
- The alkali metals form ions with ....
- Non-metals in groups 6 and 7 form ....
- The halogens form ions with ....

b Complete this table to show that ionic compounds are electrically neutral overall because the positive and negative charges balance.

Ionic compound	Ions present		Formula
	Positive ions	Negative ions	
	Li^{+}		LiBr
magnesium iodide	Mg^{2+}	I^{-} I^{-}	
			$AlBr_3$
sodium oxide			

15 Chemical hazards

Complete the descriptions of chemical hazards by filling in the blanks. Then write the names of these chemicals alongside the matching hazards. Some chemicals have more than one hazard.

chlorine	bromine	iodine	sodium	sodium hydroxide
potassium nitrate	sodium carbonate		dilute	hydrochloric acid

Symbol	Hazard	Examples
	Harmful A chemical which may involve limited health risks if breathed in, ... or taken in through the skin. (Less hazardous than a toxic chemical.)	
	Toxic A chemical which may involve serious health risks or even .... if breathed in, swallowed, or taken in through the	
	Irritant A chemical which may cause sores or inflammation in contact with skin or eyes. (Less hazardous than a .... chemical.)	
	Corrosive A chemical which can destroy living tissues such as .... or	
	Highly flammable A chemical which may easily catch .... or which gives of a flammable gas in contact with water.	
	Oxidizing A chemical which reacts strongly with other chemicals and makes the mixture so hot that it may cause a	

Chemicals of the natural environment – Higher C5

1 Chemicals in four spheres

Write the names of these chemicals in the boxes on the diagram. Some chemicals belong in more than one box.

argon	DNA	nitrogen	sodium chloride
chalk	fat	oxygen	starch
carbon dioxide	iron ore	protein	water
crude oil	granite	sandstone	

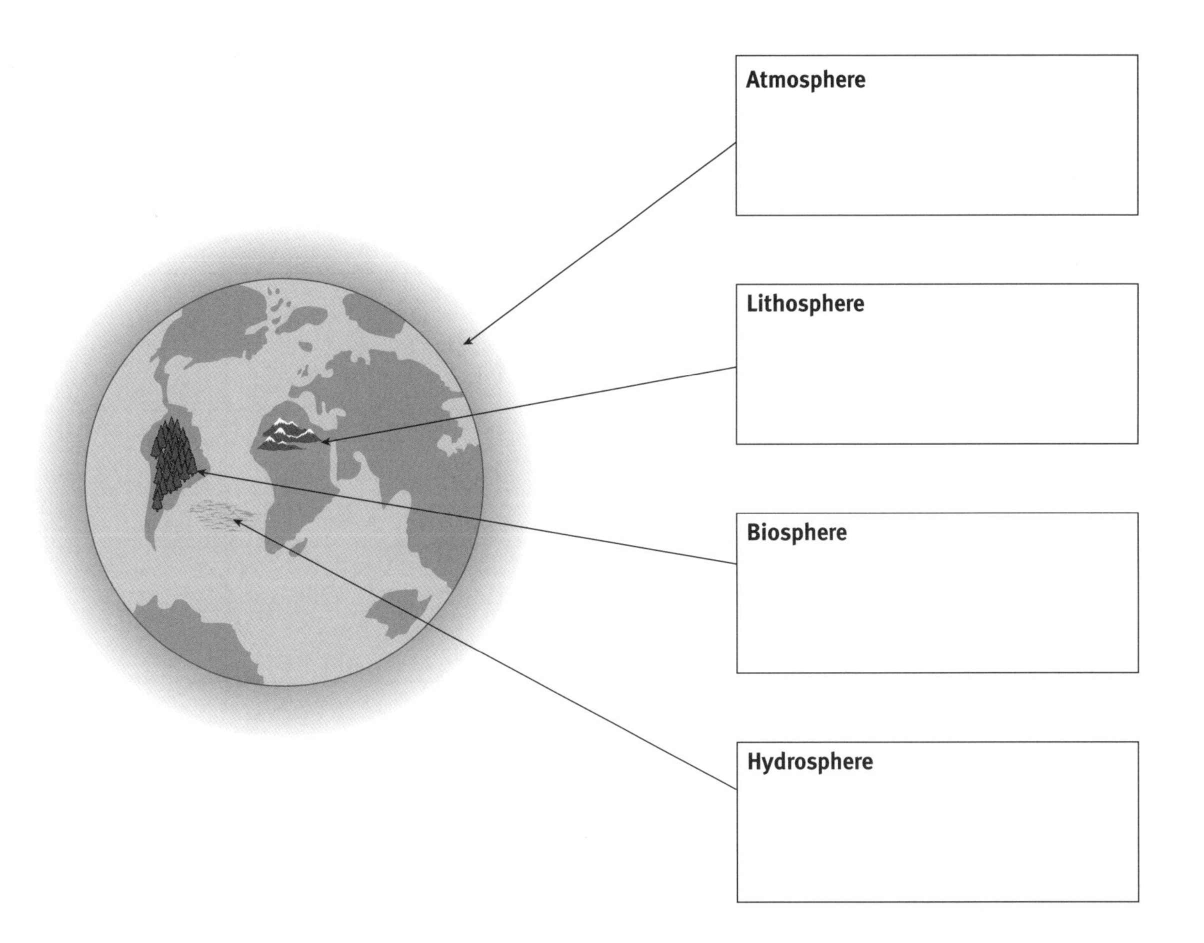

2 Gases in air

Complete this table to show the gases in unpolluted, dry air.

Gas	Element/compound	Percentage by volume in dry air
		78
	element	
argon		1
	compound	0.04

3 Chemicals of the atmosphere

Each chemical in the atmosphere consists of small molecules and is either a non-metallic element or a compound made from non-metallic elements.

a Complete this table.

Name of chemical found in air	Draw a picture of a molecule of the chemical	Write the formula that represents the molecule	Is it a non-metallic element or a compound of non-metal elements?	Is the chemical produced by human activity?
nitrogen				
oxygen				
argon				
carbon dioxide				
sulfur dioxide				
carbon monoxide				
methane				

b Complete these sentences.

- Chemicals made up of small molecules have low boiling points because ....

  ....

  ....

- Water consists of small H_2O molecules, but its boiling point is higher than molecules of comparable size and it is a liquid at normal temperatures. This is because the attractive forces between its molecules are .... than expected.

4 Strong bonds in molecules

There are strong bonds between atoms in molecules. These are covalent bonds. Colour the models and complete this table to show the bonding in molecules with the help of this information.

Atom	Usual number of covalent bonds	Colour code in models
H, hydrogen	1	white
C, carbon	4	black
O, oxygen	2	red
N, nitrogen	3	blue
Cl, chlorine	1	green

Chemical	Molecular model	Covalent bonds in the molecule	Molecular formula
hydrogen		H–H	H_2

nitrogen			
water	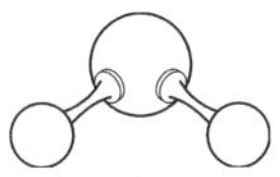		
	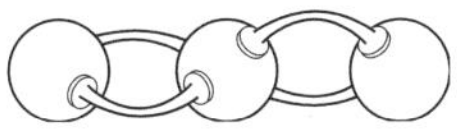		
methane	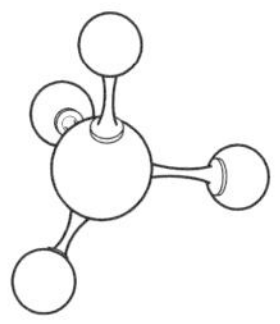		

Chemical	Molecular model	Covalent bonds in the molecule	Molecular formula
chlorine			
hydrogen chloride			
ammonia			
ethene			

5 Covalent bonds

The diagram represents how a covalent bond holds two hydrogen atoms together to make an H_2 molecule.

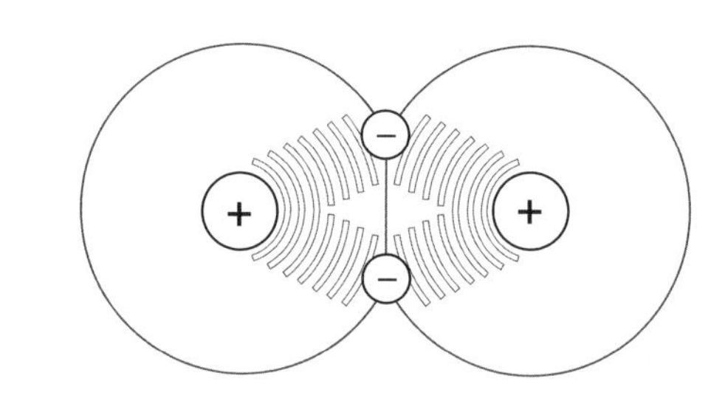

Each time a word below is printed in outline, colour it in.

Then shade the corresponding part of the diagram with the same colour.

In the hydrogen molecule (H_2) the two hydrogen atoms are held together by the electrostatic attraction between the nuclei of the two and the shared pair of electrons.

6 Properties of ionic compounds

The properties of ionic compounds are explained by their structure.

a Colour the diagram of the structure of sodium chloride. Colour the sodium ions red and the chloride ions **green**. Then complete the labels.

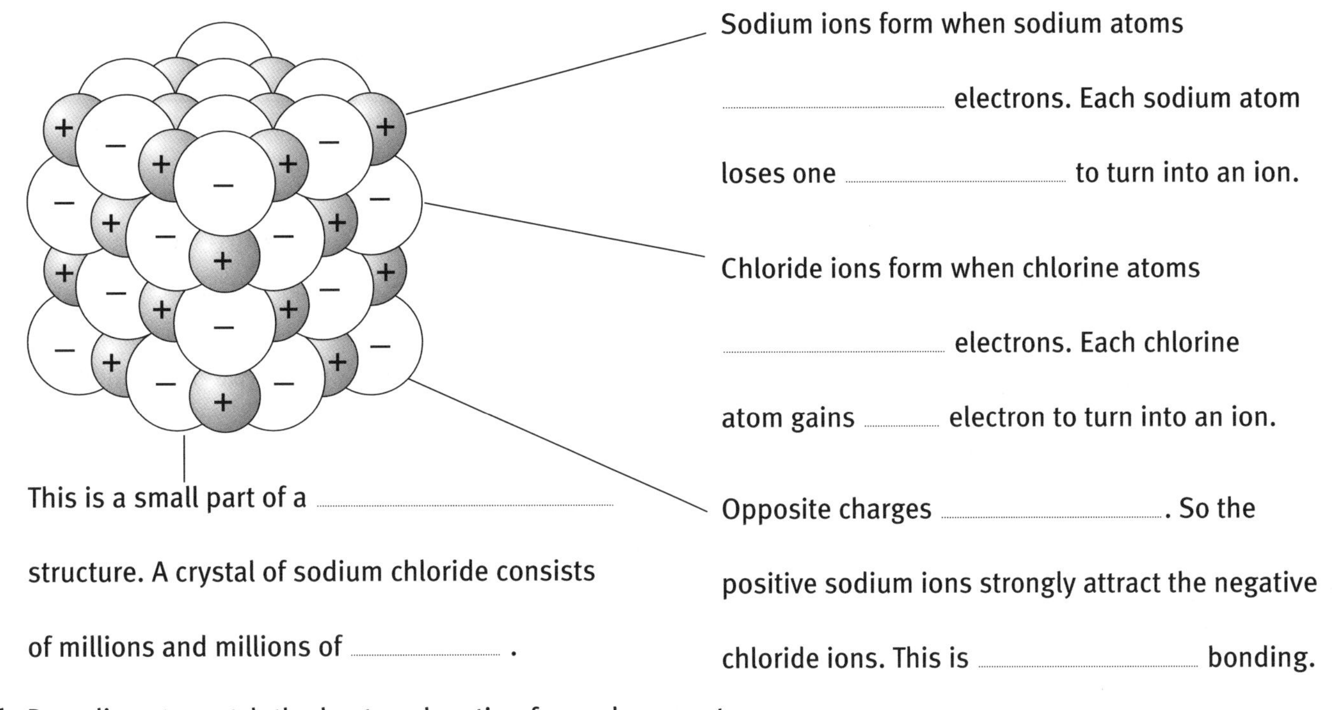

Sodium ions form when sodium atoms .... electrons. Each sodium atom loses one .... to turn into an ion.

Chloride ions form when chlorine atoms .... electrons. Each chlorine atom gains .... electron to turn into an ion.

Opposite charges .... So the positive sodium ions strongly attract the negative chloride ions. This is .... bonding.

This is a small part of a .... structure. A crystal of sodium chloride consists of millions and millions of ....

b Draw lines to match the best explanation for each property.

Property

- All the crystals of each solid ionic compound are the same shape. Whatever the size of the crystal, the angles between the faces of the crystal are always the same.
- The solution of an ionic compound in water is a good conductor of electricity.
- Ionic compounds have relatively high melting points.
- When an ionic compound is heated above its melting point, the molten compound is a good conductor of electricity.

Explanation

- The giant ionic structure is held together by the strong attraction between the positive and negative ions. It takes a lot of energy to break down the regular arrangement of ions.
- The ions in the giant ionic structure of an ionic compound are always arranged in the same regular way.
- In a molten ionic compound the positive and negative ions can move around independently.
- In a solution of an ionic compound, the positive metal ions and the negative non-metal ions can move around independently.

7 Silicon dioxide

Silicon dioxide is a common compound in the crust of the Earth.

a Complete the sentences beside the diagram of a quartz crystal.

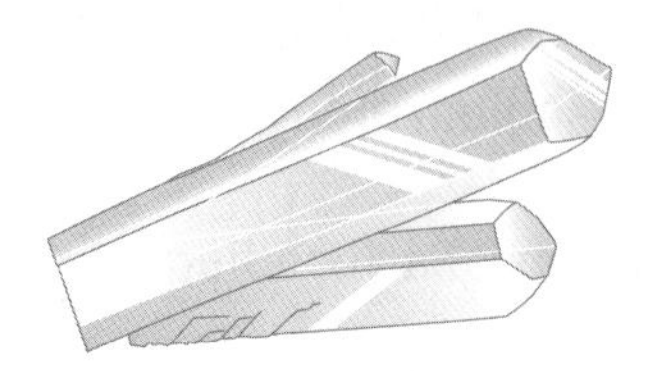

Quartz is one of the crystalline forms of ....

There are small .... crystals in granite.

The grains in sandstone consist of ....

b Colour this diagram of the structure of quartz. Colour the oxygen atoms **red** and the silicon atoms **grey**. Then complete the labels.

This is an example of a .... structure.

Each oxygen atom forms ...... covalent bonds.

Each silicon atom forms ...... covalent bonds.

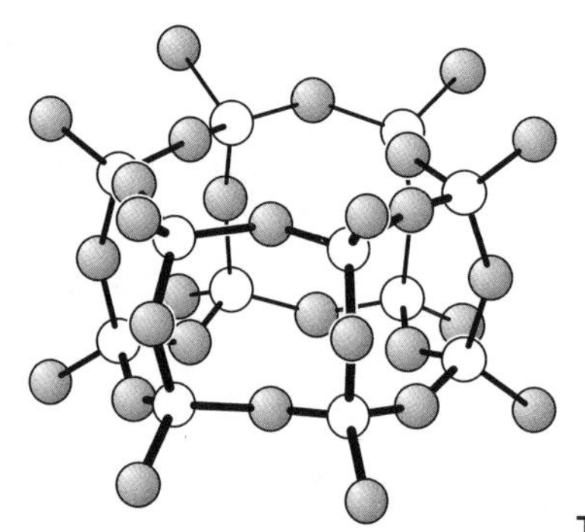

The strong bonds between the atoms are .... bonds.

There are two oxygen atoms for every silicon atom, so the formula is ....

c Complete the table describing properties and uses of silicon dioxide.

Property of silicon dioxide	Comments	Use based on the property
	scratches steel	used as an abrasive in sandpaper
high melting and boiling points		used to make furnace linings and laboratory glassware
	when granite weathers, it ends up as sand in rivers and on beaches	sandstone is used as building stone
	no free electrons in the structure to carry electricity	

8 The abundance of elements

Compare the percentages of elements in the Earth's crust and in the human body.

a Colour the diagrams using the same colour for any element that appears in both diagrams.

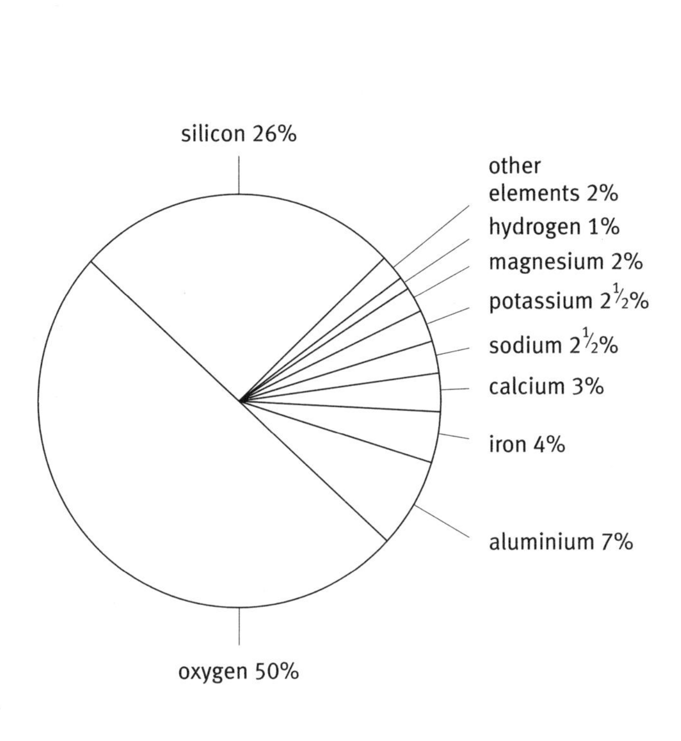

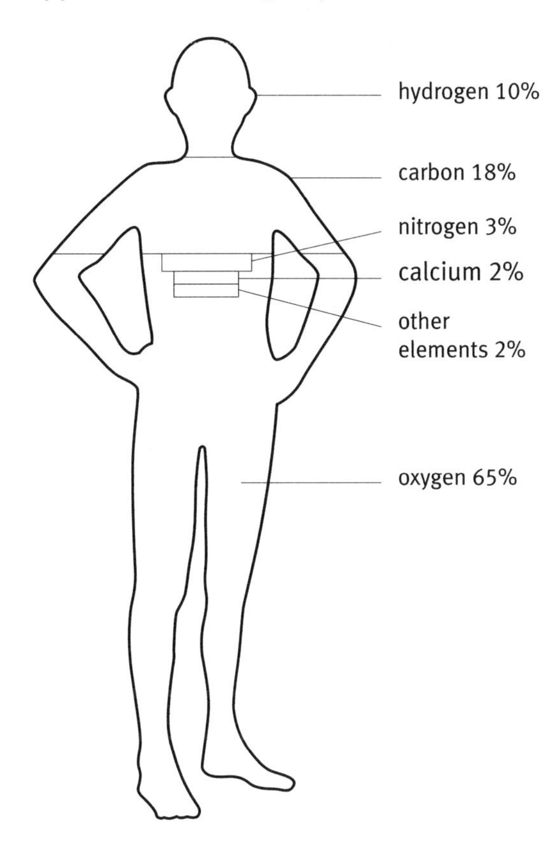

b The element which is most abundant in the Earth's crust and in the human body is ....

c Two common compounds on Earth which contain oxygen are .... and ....

d The two most abundant metals in the Earth's crust are .... and ....

e The four elements present in the largest quantities in the human body are ....

f Two 'other elements' present in the body which are important to life are .... and ....

g Molecules of carbohydrates, proteins, and DNA all include atoms of these three elements:

....

h The element present in DNA but not in carbohydrates and proteins is ....

9 Molecules in living things

Complete this table which includes information about molecules found in living things.

Molecule structure	Elements present	Formula	Type of chemical
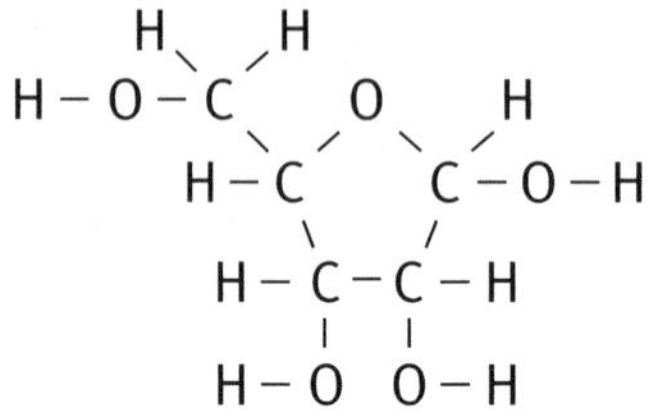 H H H – O – C O H H – C C – O – H H – C – C – H H – O O – H			sugar
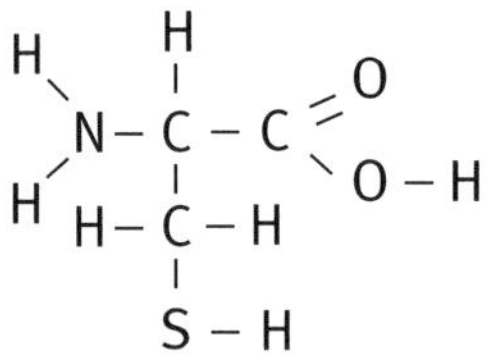 H H O N – C – C O – H H H – C – H S – H			amino acid
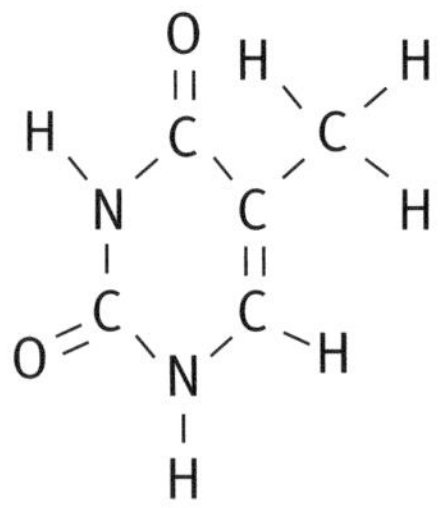 O H H H C C N C H C C H O N H			one of the bases in DNA
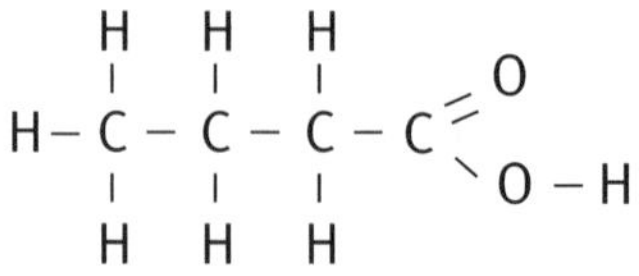 H H H O H – C – C – C – C O – H H H H			an acid found in fats

10 The oxygen cycle

Add to this diagram to show parts of the natural oxygen cycle. This cycle involves the element oxygen but also compounds of oxygen – including water, carbon dioxide, carbohydrates and so on.

Label the diagram to show parts of the:

atmosphere	hydrosphere	lithosphere	biosphere

Add arrows and labels to the diagram to show oxygen (or one of its compounds):

- moving from the biosphere to the atmosphere
- moving from the atmosphere to the biosphere
- moving from the hydrosphere to the atmosphere
- moving from the atmosphere to the hydrosphere
- moving from the atmosphere to the lithosphere

For each arrow, add a note to say what type of change is happening.

11 Metals and metal ores

Complete the questions on this page with the help of the two tables.

Table 1

Metal	**Metal ore**	**Formula of the mineral in the metal ore**
aluminium	bauxite	Al_2O_3
iron	magnetite	Fe_3O_4
potassium	sylvite	KCl
tin	cassiterite	SnO_2
zinc	zincite	ZnO_2

Table 2

Reactivity	**Metal**
most reactive	Na
	Al
The more reactive a metal is, the more strongly it holds onto oxygen and the more difficult it is to extract the metal.	Zn
	Fe
	Pb
	Cu
	Ag
least reactive	Au

a Name an element from table **2** that that can be found free in nature. Why is this element found uncombined with other elements?

....

b Name three metals in table **1** that can be extracted from their oxide ores by heating with carbon.

....

c Add the missing state symbols and balance this equation that shows the use of carbon to extract a metal from an ore.

$Fe_3O_4(s)$ + C(.....) → Fe(s) + CO_2(.....)

This equation shows the extraction of the metal iron.

The chemical that is reduced is ....

The chemical that is oxidized is ....

d Name two elements in table **1** that cannot be extracted by heating with carbon. Give a reason for your choice.

....

....

Name the method used to extract these metals ....

e Explain why it is sometimes necessary to mine large amounts of waste rock when extracting a metal such as copper.

....

....

12 Calculating percentages of metal in metal ores

a Complete this diagram to work out the formula mass of the iron oxide in the ore magnetite.

Relative atomic masses: Fe = 56	O = 16

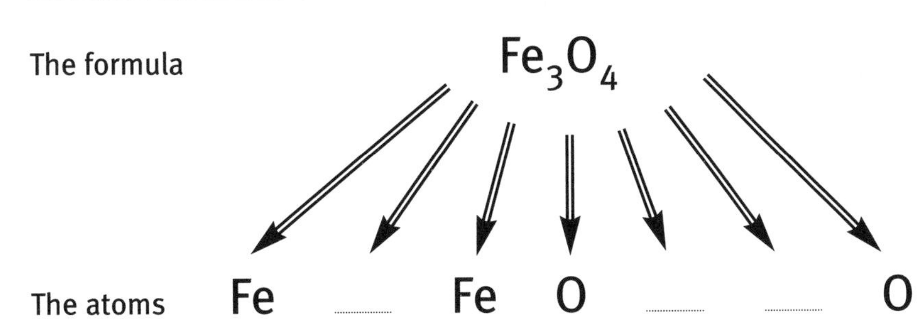

The relative atomic masses ....

The relative formula mass of the iron oxide = ...

In this formula there are .... atoms of iron, Fe.

The relative mass of .... Fe = ...

This means that in .... kg of Fe_3O_4 there are .... kg of Fe.

So 1 kg of Fe_3O_4 contains .... kg of Fe

So 100 kg of Fe_3O_4 contains .... kg of Fe

Another way of saying this is that the percentage of Fe in Fe_3O_4 = ... %

b Work out the percentage of copper in bornite by filling in the gaps to show your working.

Relative atomic masses: Cu = 64	O = 16	S = 32

Bornite, Cu_5FeS_4, formula mass = ...

Relative mass of copper in the relative formula mass = ...

This means that in .... kg of Cu_5FeS_4 there are .... kg of Fe.

So 1 kg of Cu_5FeS_4 contains .... kg of Cu

So 100 kg of Cu_5FeS_4 contains .... kg Cu

So the percentage of Cu in Cu_5FeS_4 = ... %

13 Electrolysis of sodium chloride

a Complete the labelling of the diagram to show what happens during the electrolysis of molten sodium chloride. Choose from these words.

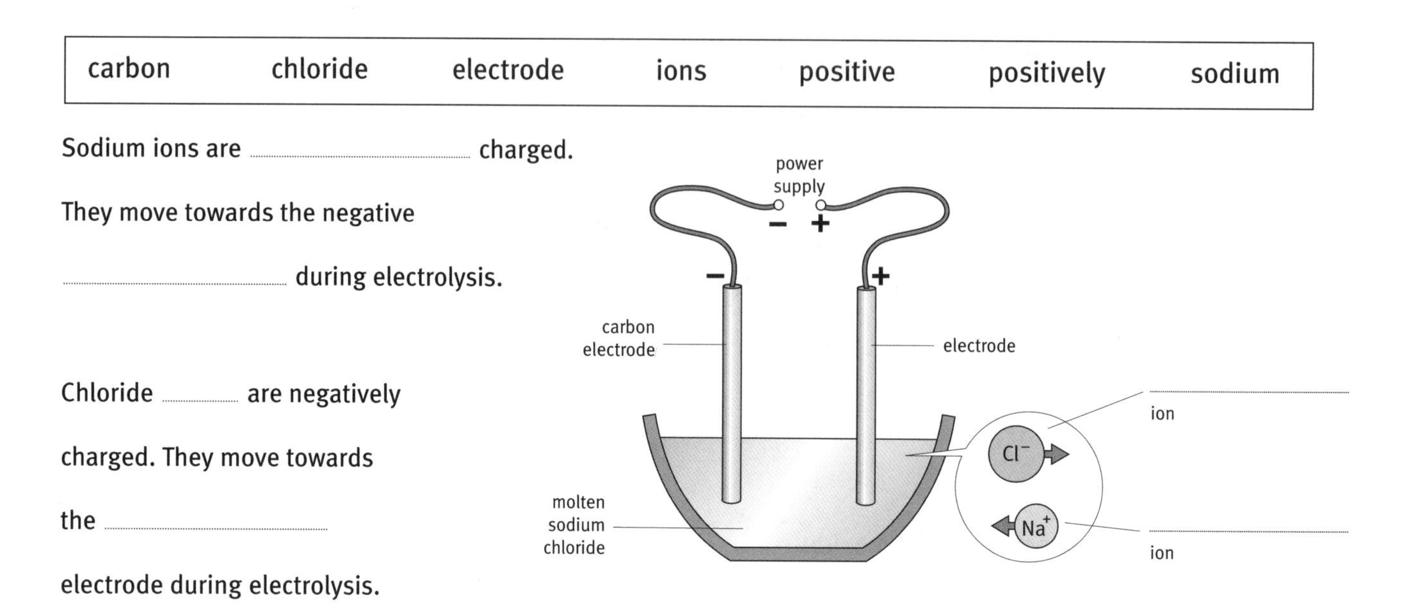

carbon	chloride	electrode	ions	positive	positively	sodium

Sodium ions are .... charged.

They move towards the negative .... during electrolysis.

Chloride .... are negatively charged. They move towards the .... electrode during electrolysis.

b Complete the labelling of the diagrams to explain what happens during the electrolysis of molten sodium chloride. Choose from these words.

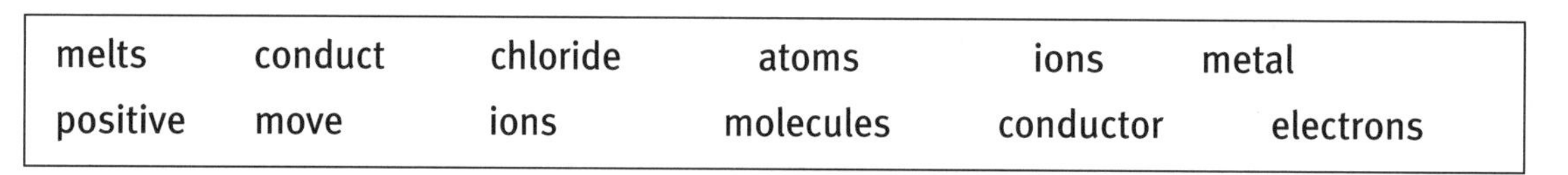

melts	conduct	chloride	atoms	ions	metal
positive	move	ions	molecules	conductor	electrons

The .... in solid sodium chloride cannot move around, so salt does not .... electricity.

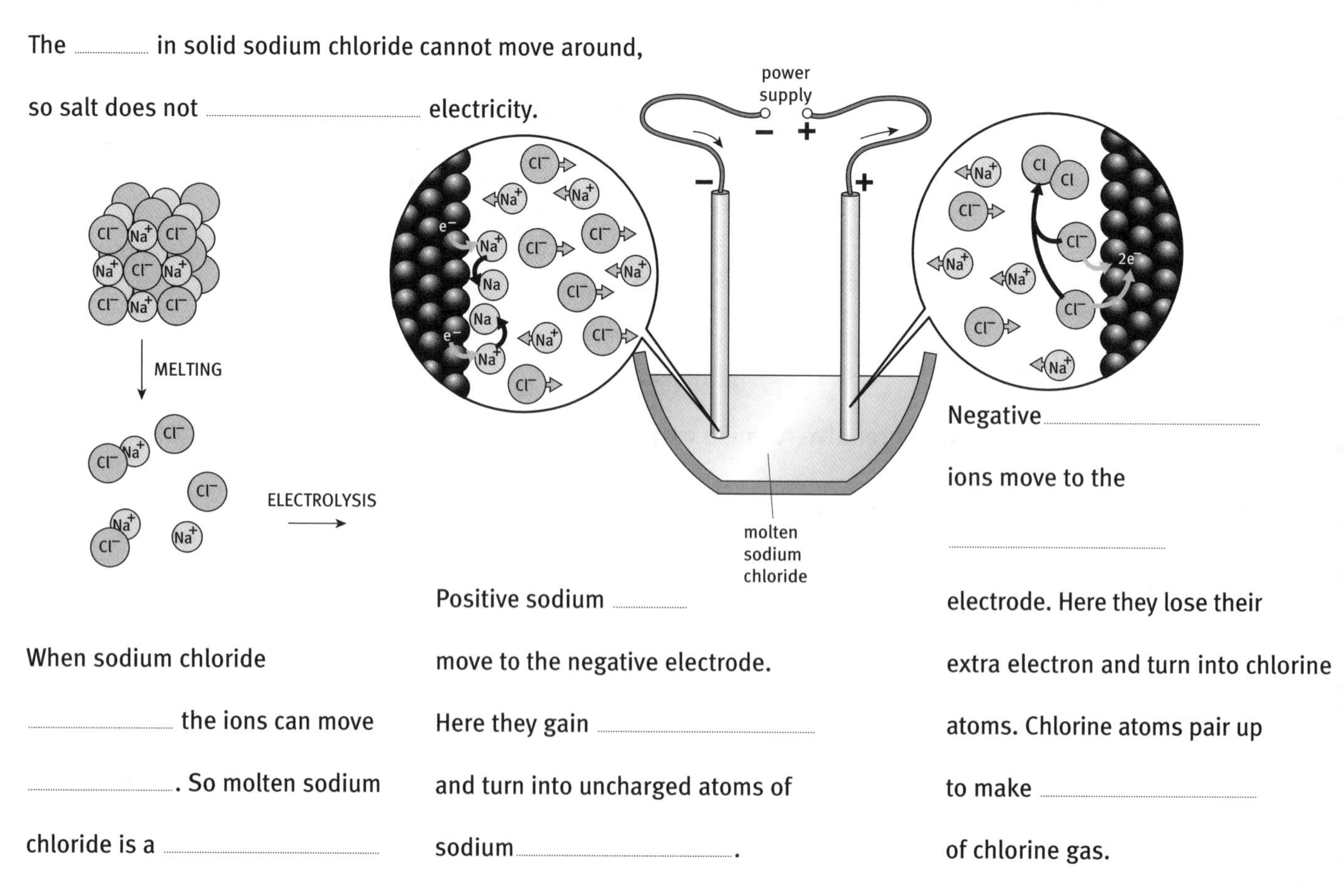

When sodium chloride .... the ions can move .... So molten sodium chloride is a ....

Positive sodium .... move to the negative electrode. Here they gain .... and turn into uncharged atoms of sodium ....

Negative .... ions move to the .... electrode. Here they lose their extra electron and turn into chlorine atoms. Chlorine atoms pair up to make .... of chlorine gas.

14 Electrolysis of aluminium oxide

a Label the diagram to describe the equipment used to extract aluminium from aluminium oxide. Use these words and phrases.

carbon anodes	carbon lining	negative electrode
molten aluminium oxide	molten aluminium	tapping hole

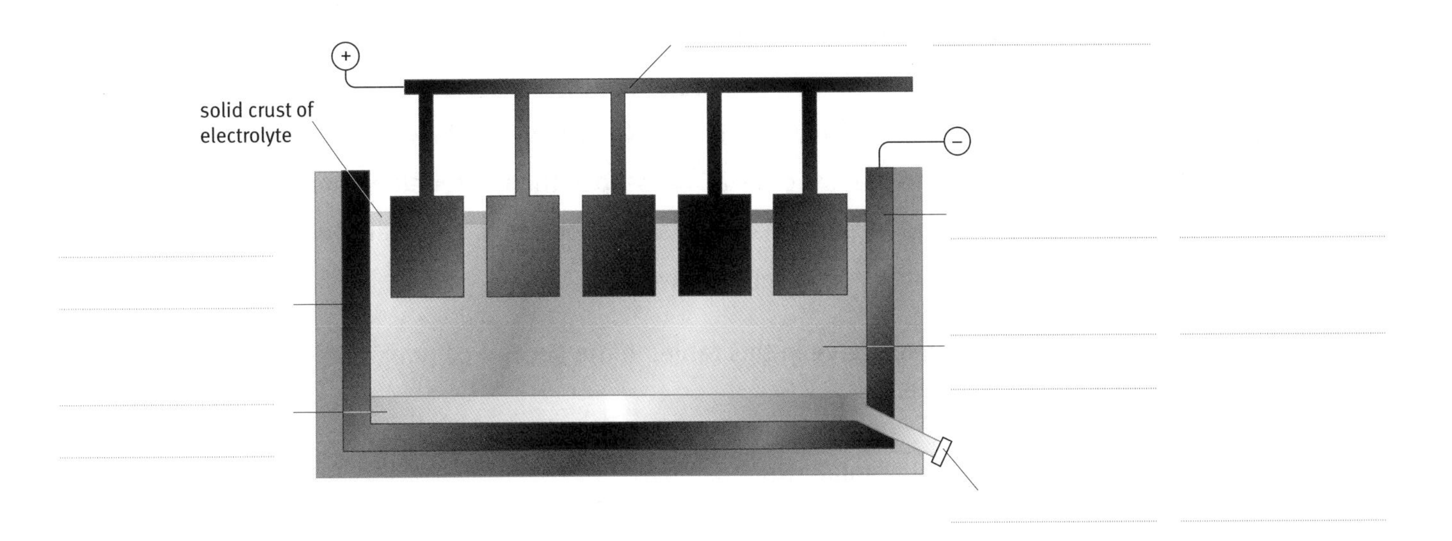

b Explain why aluminium oxide conducts electricity only when liquid but not when solid.

....

....

c Write the symbols for the two ions in aluminium oxide: ....

d Complete this equation to show what happens at the negative electrode during the electrolysis of molten aluminium oxide.

.... $+ 3e^- \rightarrow Al$

e Complete these equations to show what happens at the positive electrode during the electrolysis of molten aluminium oxide.

.... $\rightarrow O + 2e^-$

.... + ... $\rightarrow O_2$

15 Properties of metals

a Read the following paragraph and underline four different properties of metals.

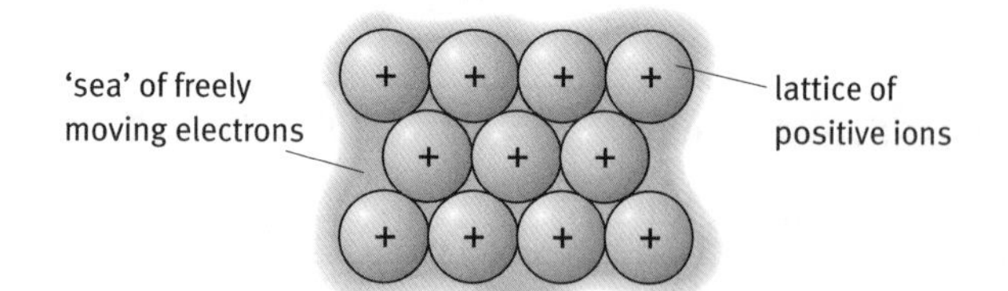

A model of metallic bonding

In a metal, such as copper, the atoms are packed closely together. The atoms are held together by strong metallic bonds, so copper is strong and difficult to melt. Copper is malleable, which means that it can be beaten into a different shape because the atoms can be moved around without the structure losing its strength. When a metal structure is formed, the metal atoms lose their outer electrons and form positive ions. The electrons are no longer held by particular atoms, so they can move freely between the positive ions. This is why metals are good conductors of electricity.

b List the four properties you underlined above in this table. In the second column, make notes on how metallic bonding explains each property.

Property	Explanation
1	
2	
3	
4	

16 The life cycle of a metal

Use these words and phrases to complete the diagram which shows the life cycle of a metal. Two of the boxes have been filled in for you.

metal in use	rubbish to waste tip	mining the ore
separating and purifying the mineral with the metal	separating and recycling waste metal	recycling scrap metal
making products from the metal	extracting the metal from the mineral	shaping the metal into sheets, wire, or bars
end of useful life		

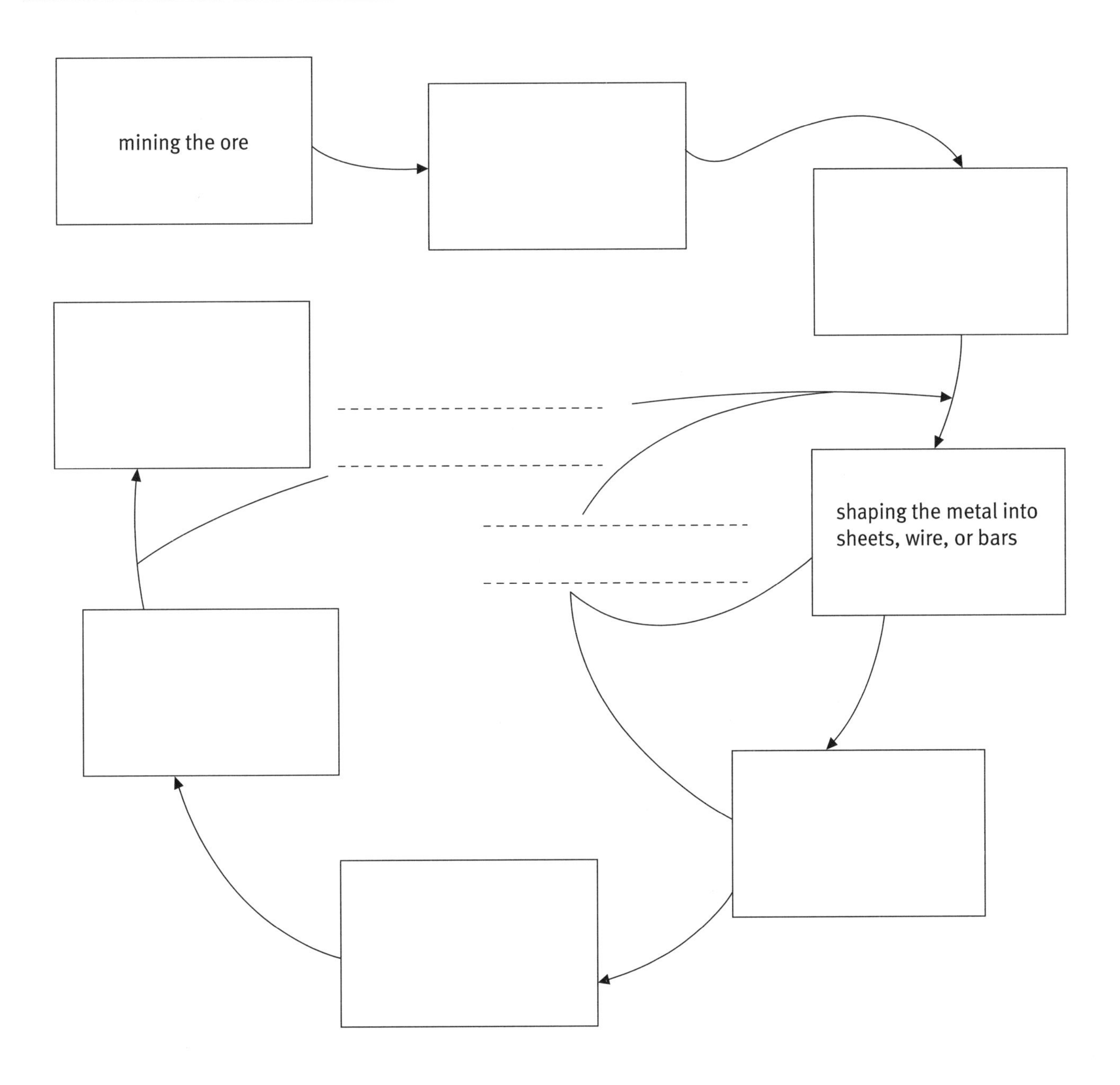

Chemical synthesis – Higher C6

1 The chemical industry

This chart shows the value of the sales of the various sectors of the chemical industry in the EU. The figures are for 2004.

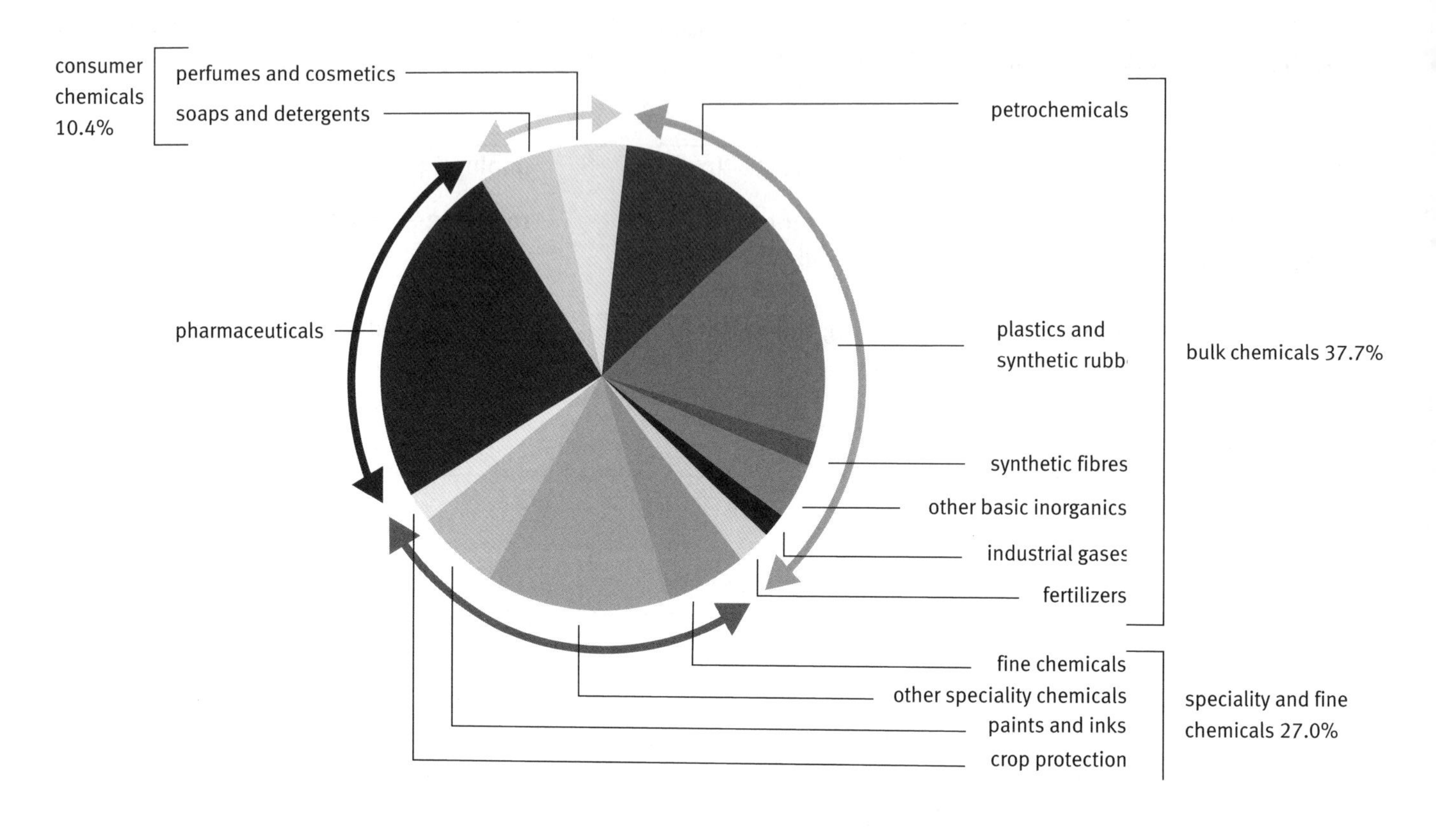

a What is the raw material for making petrochemicals? ....

b Give an example of a petrochemical that is used to make plastics on a large scale. ....

c Give an example of a type of chemical needed for 'crop protection'. ....

d Why are paints and inks now so valuable as products? ....

....

e Give the chemical formulae of the molecules of these three gases that are used in industry:

- nitrogen ....
- oxygen ....
- hydrogen ....

f What percentage value of products of the EU chemical industry are used:

- as perfumes, cosmetics, soaps, and detergents? ....
- for diagnosis and treatment in medicine? ....

2 Acids and alkalis

a Complete this table about some acids.

Name of acidic compound	**Formula of acidic compound**	**State of the pure compound at room temperature**
citric acid	$C_6H_8O_7$	
tartaric acid	$C_4H_6O_6$	
	H_2SO_4	
	HNO_3	
ethanoic acid	CH_3COOH	
hydrogen chloride		

b Complete this table about some alkalis.

Name of alkaline compound	**Formula of alkaline compound**	**State of the pure compound at room temperature**
	NaOH	
potassium hydroxide		
calcium hydroxide		solid

c Acids and alkalis show their characteristic reactions when dissolved in water. Draw a line to match each of these solutions to its pH value.

pure water	pH 14
dilute hydrochloric acid	pH 12
dilute sodium hydroxide solution	pH 7
vinegar	pH 3
limewater (calcium hydroxide solution)	pH 1

3 Reactions of acids

- Complete the general word equations to summarize the main reactions of solutions of acids in water.
- Next complete the word equations for the examples.
- Finally complete and balance the matching symbol equations, including state symbols.

a Acids with metals

The general word equation for the reaction of an acid with a metal is:

acid + ... → salt + hydrogen

Example

.... + magnesium → magnesium chloride + hydrogen

$HCl(\ldots)$ + ... → $MgCl_2(\ldots)$ + ...$(\ldots)$

b Acids with metal oxides

The general word equation for the reaction of an acid with a metal oxide is:

acid + metal oxide → salt + ...

Example

nitric acid + copper oxide → .... + ...

............$(\ldots)$ + $CuO(\ldots)$ → $Cu(NO_3)_2(\ldots)$ + ...$(\ldots)$

c Acids with metal hydroxides

The general word equation for the reaction of an acid with a metal hydroxide is:

acid + metal hydroxide → .... + water

Example

.... + ... → sodium sulfate + ...

$H_2SO_4(\ldots)$ + $NaOH(\ldots)$ → ....$(\ldots)$ + $H_2O(\ldots)$

d Acids with metal carbonates

The general word equation for the reaction of an acid with a metal carbonate is:

acid + metal carbonate → .... + carbon dioxide + water

Example

.... + ... → calcium chloride + carbon dioxide + ...

$HCl(\ldots)$ + $CaCO_3(\ldots)$ → ....$(\ldots)$ + ...$(\ldots)$ + $H_2O(\ldots)$

4 Ions and formulae

The table shows the charges on common ions found in ionic compounds.

a Complete the table

Positive ions			**Negative ions**		
Ion	**Charge**	**Symbol**	**Ion**	**Charge**	**Symbol**
lithium	....	Li^{+}	chloride	1^{-}	....
sodium	$1+$	....	....	1^{-}	Br^{-}
....	$1+$	K^{+}	iodide	....	I^{-}
magnesium	$2+$	....	nitrate	1^{-}	NO_3^{-}
....	....	Ca^{2+}	....	1^{-}	OH^{-}
barium	$2+$	Ba^{2+}	oxide	2^{-}	....
....	$3+$	Al^{3+}	....	....	CO_3^{2-}
			sulfate	....	SO_4^{2-}

b Use the complete table to help you to write down the formulae of these ionic compounds.

sodium hydroxide ....

sodium chloride ....

sodium nitrate ....

sodium carbonate ....

potassium chloride ....

magnesium bromide ....

magnesium oxide ....

magnesium hydroxide ....

magnesium carbonate ....

magnesium sulfate ....

calcium carbonate ....

calcium chloride ....

calcium iodide ....

calcium nitrate ....

aluminium oxide ....

aluminium chloride ....

c Work out the symbol for the ions in these compounds that are not included in the table.

- sulfide ion in barium sulfide, BaS ....
- strontium ion in strontium nitrate, $Sr(NO_3)_2$ ....
- phosphate ion in sodium phosphate, Na_3PO_4 ....

5 Testing the purity of citric acid

Procedure

(1) A 0.48 g sample of citric acid was dissolved in 50 cm^3 of water.

(2) 1 cm^3 phenolphthalein indicator was added.

(3) The solution was titrated with a solution of sodium hydroxide. These were the burette readings from the titration:

	Titration
Second burette reading/cm^3 First burette reading/cm^3	22.40 3.30
Volume of NaOH(aq) added/cm^3	

a Label these diagrams to summarize the procedure.

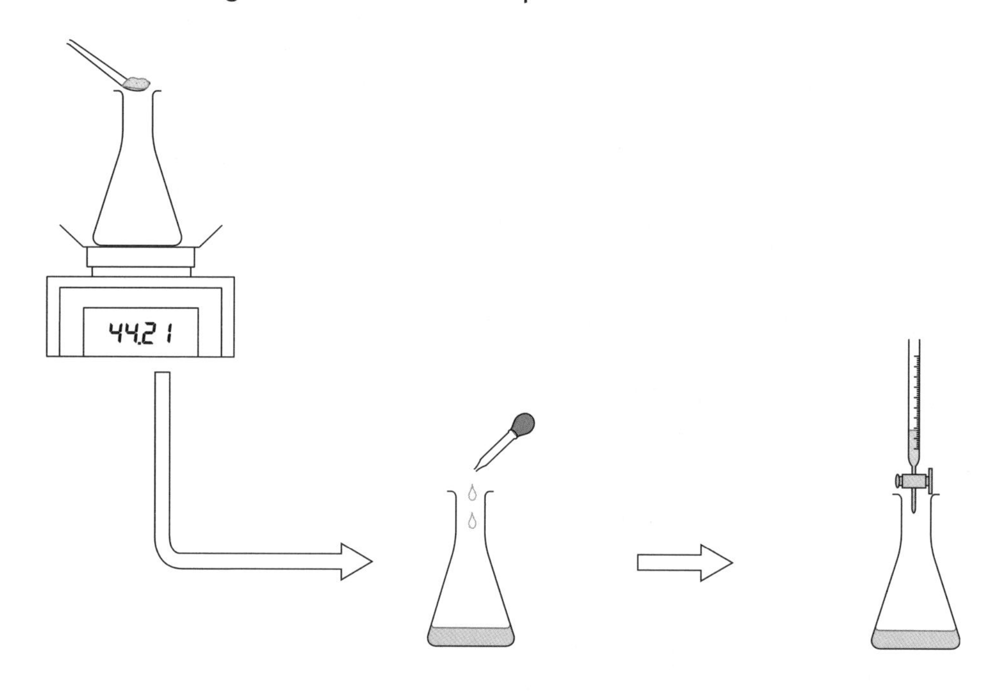

b Calculate the purity of the citric acid from the formula given below.
T is the titre (the volume of solution added from the burette).
The concentrations of the sodium hydroxide used in the titration meant that the value of $F = 0.025$.

$$\text{purity} = \frac{T \times F \times 100}{\text{mass of sample}} \%$$

purity = ... %

c Give an example to show why it is important to be able to measure the purity of a chemical such as citric acid.

....

....

6 Effect of surface area on the rate of a reaction

The diagrams show the apparatus for investigating the rate of reaction of marble chips with acid.

- Diagram **A** shows the apparatus before the reaction starts.
- Diagram **B** shows the reaction in progress.

The reaction was carried out twice – first with larger chips of marble, then with smaller chips. There were still unchanged marble chips in both flasks when the reaction stopped.

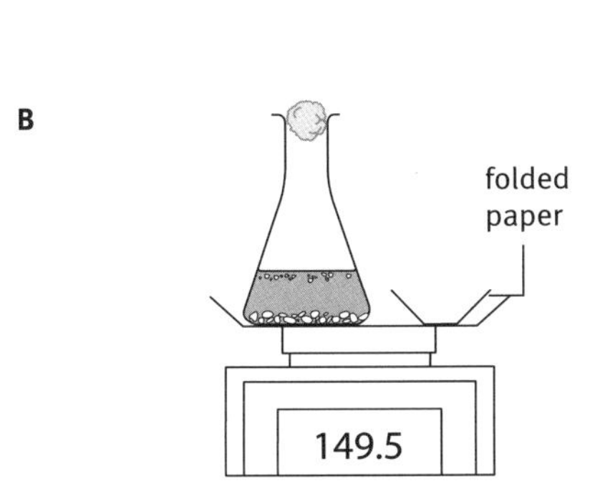

The graph shows typical results using two sizes of marble chips.

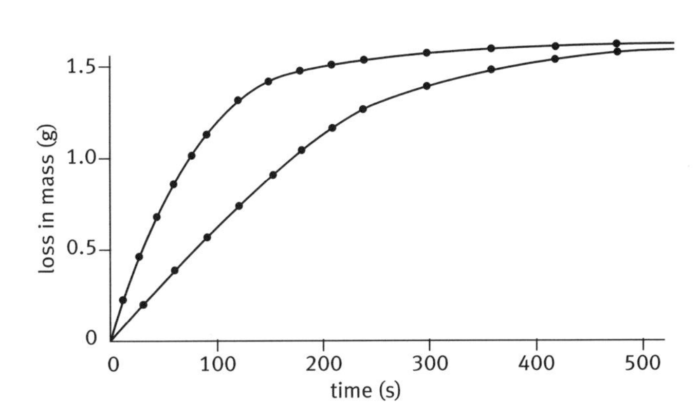

a Label the graph lines to show:

- which is the line for larger chips and which the line for smaller chips
- where the reaction was fastest
- where the reaction was slowing down
- where the reaction had stopped.

b Explain why the total mass of the flask, acid and marble fell during the reaction.

....

....

c Explain why the reaction slowed down and stopped with the same final loss in mass for both the larger chips and the smaller chips.

....

....

....

d Explain the difference in the rate of reaction at the start for the two sizes of marble chips.

....

....

....

7 Effect of concentration on the rate of reaction

Adding dilute hydrochloric acid to a solution of sodium thiosulfate starts a slow reaction.

$Na_2S_2O_3(aq) + 2HCl(aq) \rightarrow 2NaCl(aq) + H_2O(l) + SO_2(aq) + S(s)$

The mixture turns cloudy. In time it is not possible to see through the solution.

The diagram shows a method for investigating the rate of reaction. The experimenter looks down at the cross on the paper from above and records the time it takes for the cross to vanish after adding the acid.

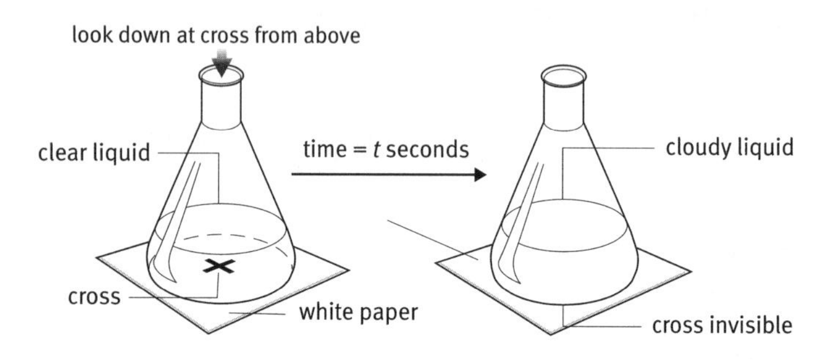

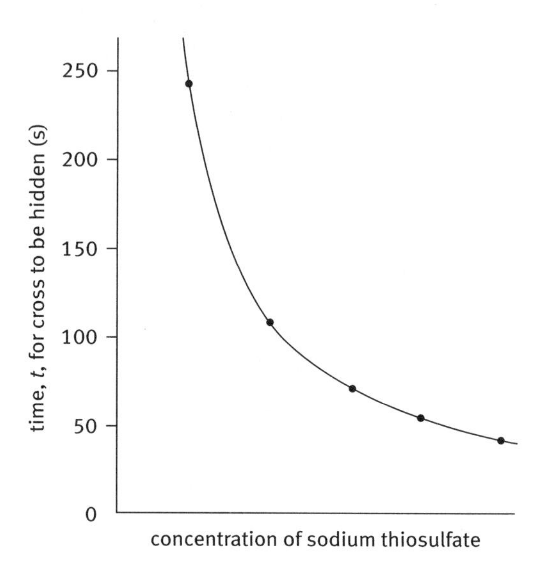

In an investigation, an experimenter added 5 cm³ of dilute hydrochloric acid to 50 cm³ samples of sodium thiosulfate solution. The experimenter measured the time for the cross to disappear with five different concentrations of sodium thiosulfate solution. The graph is a plot of the results.

a Look at the equation above and use it to explain why the solution of sodium thiosulfate turned cloudy after the hydrochloric acid was added.

....

....

b Why did the experimenter use the same volume and concentration of dilute hydrochloric acid with each different concentration of sodium thiosulfate solution?

....

....

c Put into words what the graph shows about the effect of changing the concentration of the sodium thiosulfate solution on the rate of the reaction.

....

....

....

8 The effect of temperature changes on the rate of reaction

The reaction of sodium thiosulfate with hydrochloric acid (see question 7) can also be used to investigate the effect of temperature on the rate of a reaction. In this investigation, the volume and concentration of the acid and sodium thiosulfate stay the same. The thiosulfate solution is warmed before adding the acid. The experimenter measures the temperature of the mixture after adding the acid and the time taken for the cross to disappear.

Temperature (°C)	20	30	40	50	60
Time taken for the cross to disappear (s)	280	132	59	31	17

Explain what you can conclude from the results in the table.

....

....

....

9 Predicting rates of reaction

The graph shows the volume of hydrogen produced when excess zinc granules react with 50 cm^3 dilute hydrochloric acid at 20 °C, plotted against time.

Show the effect of each change to the conditions by completing this table and also by drawing on the graph the line you would expect.

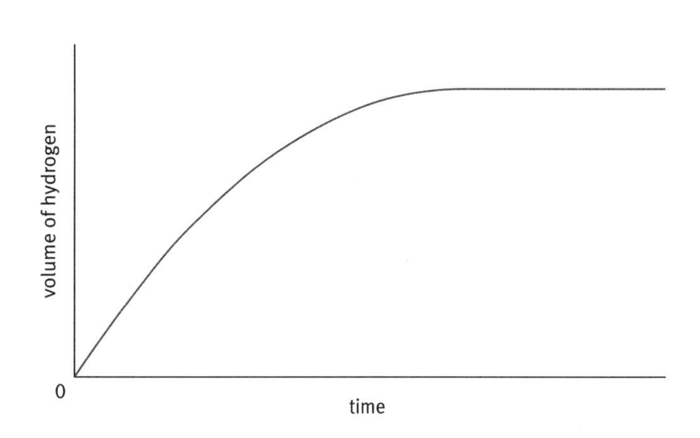

Change of conditions (all other factors stay the same)	**Effect on the rate of reaction at the start**	**Effect on the volume of gas collected (at room temperature) when the reaction stops**
halving the concentration of the acid		
carrying out the reaction at 30 °C		
using the same mass of zinc but in larger pieces		

10 The effect of catalysts on rates of reaction

Hydrogen peroxide solution contains the compound H_2O_2.

At room temperature it decomposes very slowly to give water and oxygen.

$2H_2O_2(aq) \rightarrow 2H_2O(l) + O_2(g)$

The graph shows the results from three tests. Each time a small amount of a metal oxide was added to 50 cm^3 of a solution of hydrogen peroxide in a flask. The oxygen gas was collected and measured for up to 5 minutes.

In a control experiment, with no added metal oxide, no oxygen was collected in 5 minutes.

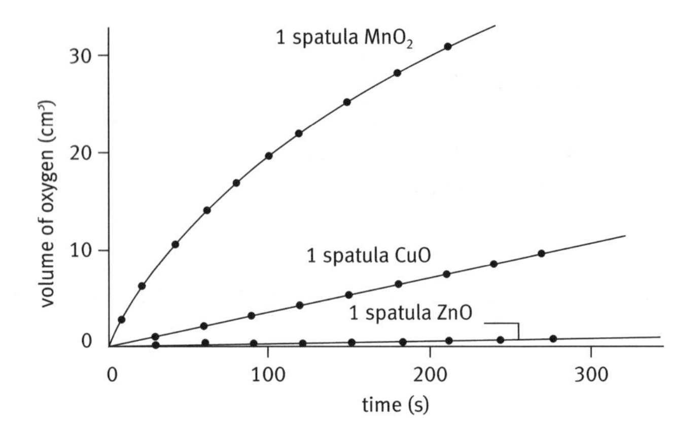

a Complete and label this diagram to show how the oxygen could be collected and measured.

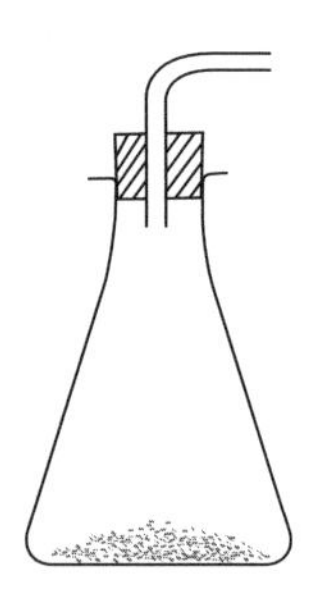

b Put the three metal oxides tested in order of their effectiveness as catalysts for the reaction.

....

c Why was a control experiment carried out?

....

d Explain the meaning of the term 'catalyst'.

....

....

....

11 Collision theory

Chemists use collision theory to explain why factors such as surface area, concentration, temperature, and catalysts affect the rates of reactions.

a Write a paragraph of four or five sentences to explain the key ideas of collision theory.

....

....

....

....

....

....

....

b Colour and label this diagram. Then use it in an explanation to show how collision theory explains why reactions in solution go faster if the concentrations of reactants are higher.

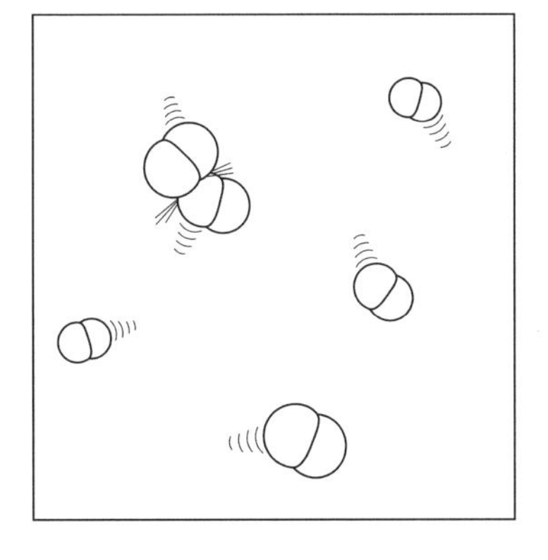

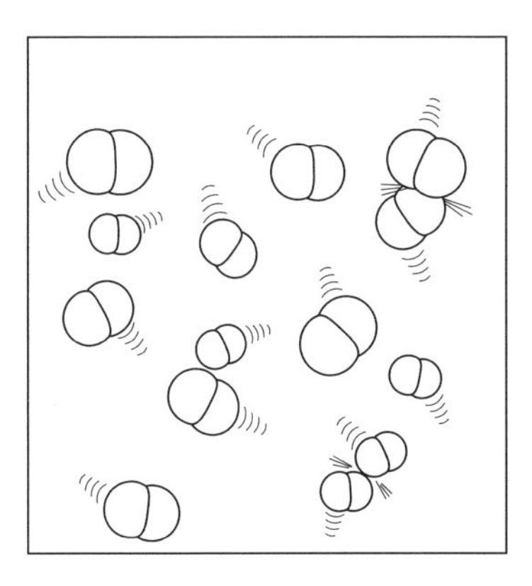

Explanation: ....

....

....

....

....

....

....

12 Making a soluble salt from an acid

The diagrams show how to make a pure sample of magnesium sulfate from magnesium oxide.

Label the diagrams. Add a caption to each stage of the diagram to describe what is happening and to explain the purpose of the stage.

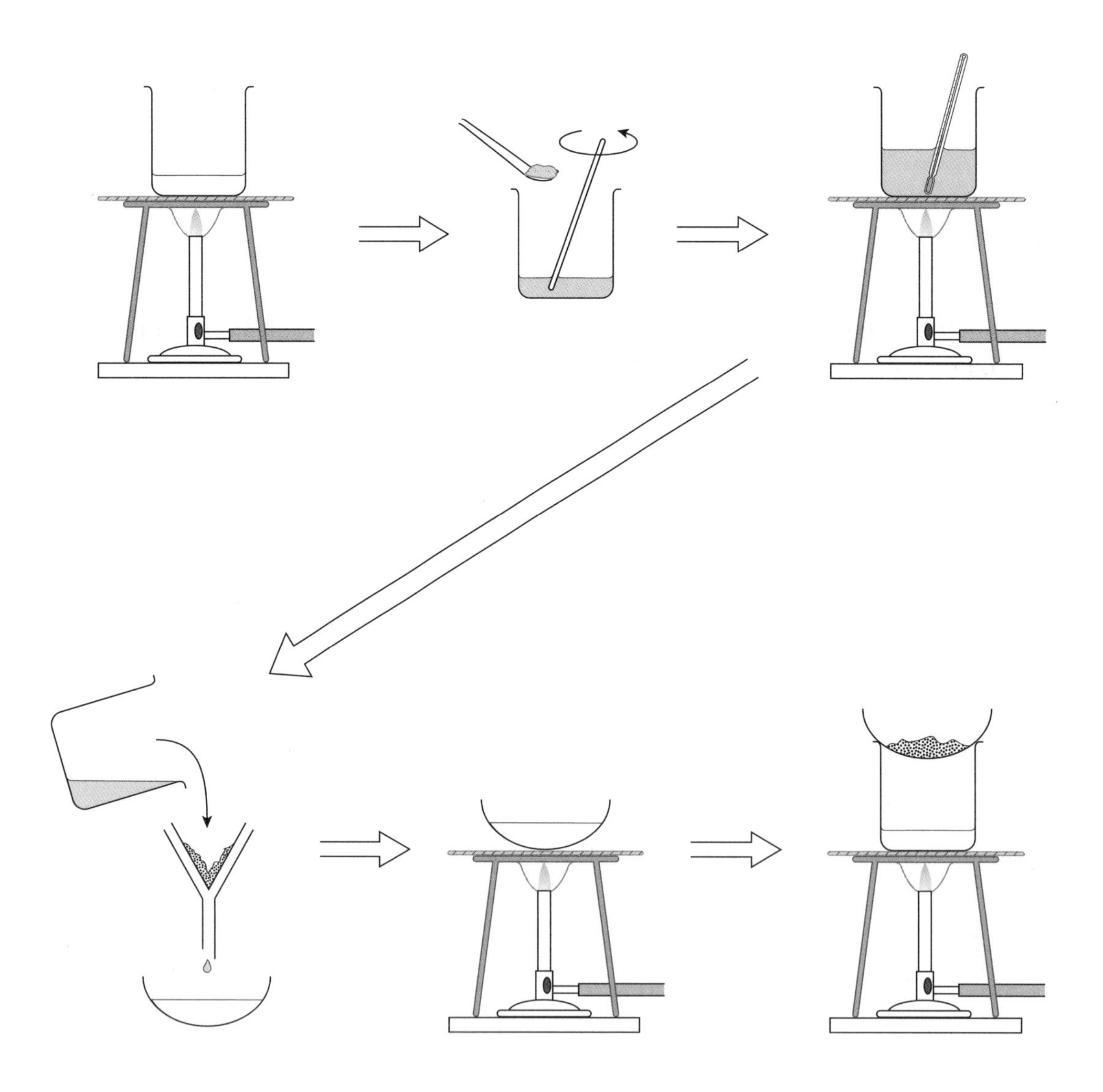

13 Stages in synthesis

Use the example on page 88 (or any other example) to explain the importance of these stages in any chemical synthesis.

a Choosing the reaction

b Working out the quantities to use

c Carrying out the reaction in suitable apparatus under the right conditions

d Separating the product from the reaction mixture

e Purifying the product

f Measuring the yield and checking the purity of the product

14 Yields

a Work out the theoretical yield of magnesium sulfate, $MgSO_4$, that can be made from 4.0 g of magnesium oxide, MgO, and an excess of sulfuric acid.

Step 1 Write the balanced symbol equation for the reaction. (Write it on the top line only.)

Step 2 Work out the formula masses of the chemicals.

Relative atomic masses: Mg = 24, O = 16, S = 32

Step 3 Write the relative reacting masses for the relevant chemicals under the balanced equation in step 1.

Step 4 Convert to reacting masses by adding the units.

Step 5 Scale the quantities to the amounts actually used to find the theoretical yield.

b What is the percentage yield if the actual yield is 10.0 g?

c Use the same steps to work out the theoretical yield of zinc sulfate, $ZnSO_4$, that can be made from 9.3 g zinc carbonate, $ZnCO_3$, and an excess of sulfuric acid. (Relative atomic masses: Zn = 65, C = 12, S = 32, O = 16.)

d What is the percentage yield if the actual yield is 11.4 g?

15 Neutralization

a Draw a line to match each statement on the left with the related box on the right.

The salt formed when sodium hydroxide reacts with sulfuric acid	NaOH
The salt formed when potassium hydroxide reacts with citric acid	soluble metal hydroxide
The acid which reacts with calcium hydroxide to form calcium nitrate	neutralization
The alkali which reacts with acetic acid to form sodium acetate	sodium sulfate
A set of compounds that are alkalis in water	HNO_3
The type of reaction that occurs when an acid and an alkali form a salt	potassium citrate

b The graph shows how the pH of a solution of hydrochloric acid changed with the volume of dilute sodium hydroxide added.

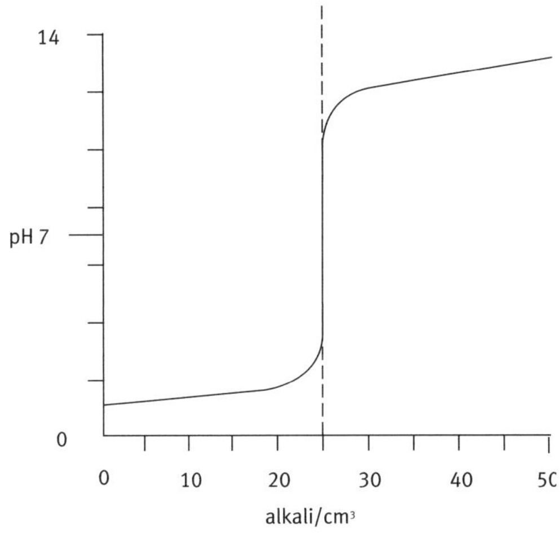

14
pH 7
0
0 10 20 30 40 50
alkali/cm³

➭ What volume of the dilute sodium hydroxide solution was needed to neutralize the dilute acid?

Explain your answer.

....

....

....

....

➭ The neutralization can be carried out with a few drops of universal indicator in the solution.
Use colours to indicate on the graph the colours you would expect to see at each stage as the acid is added to the alkali.

....

....

....

....

16 Ionic theory of neutralization

a Use these words and symbols to complete the text.

acid	alkali	hydrogen	hydroxide	ions	molecules	salt	sodium
$H^+(aq)$	$H_2O(l)$	$H_2O(l)$	$K^+(aq)$	$NO_3^-(aq)$	$OH^-(aq)$		

- Acids are chemicals containing .... which react with water to give hydrogen .... in solution.

  $HNO_3(l)$ + water → .... + $NO_3^-(aq)$

- Alkalis are ionic compounds. Examples are the soluble hydroxides of the alkali metals (lithium, ..., and potassium). These compounds consist of metal ions and .... ions. When they dissolve they add hydroxide ions to water.

  $KOH(aq) \rightarrow K^+(aq) +$ ....

- Potassium hydroxide and nitric acid react to make a .... (potassium nitrate) and water.

  .... $+ OH^-(aq) + H^+(aq) +$ .... $\rightarrow K^+(aq) + NO_3^-(aq) +$ ....

- During a neutralization reaction the hydrogen ions from the .... react with the hydroxide ions from the .... to make water ....

  $H^+(aq) + OH^-(aq) \rightarrow$ ....

  The remaining ions in solution make a salt.

b The diagram illustrates the changes to ions when dilute hydrochloric acid reacts with dilute sodium hydroxide solution. Complete the diagram.

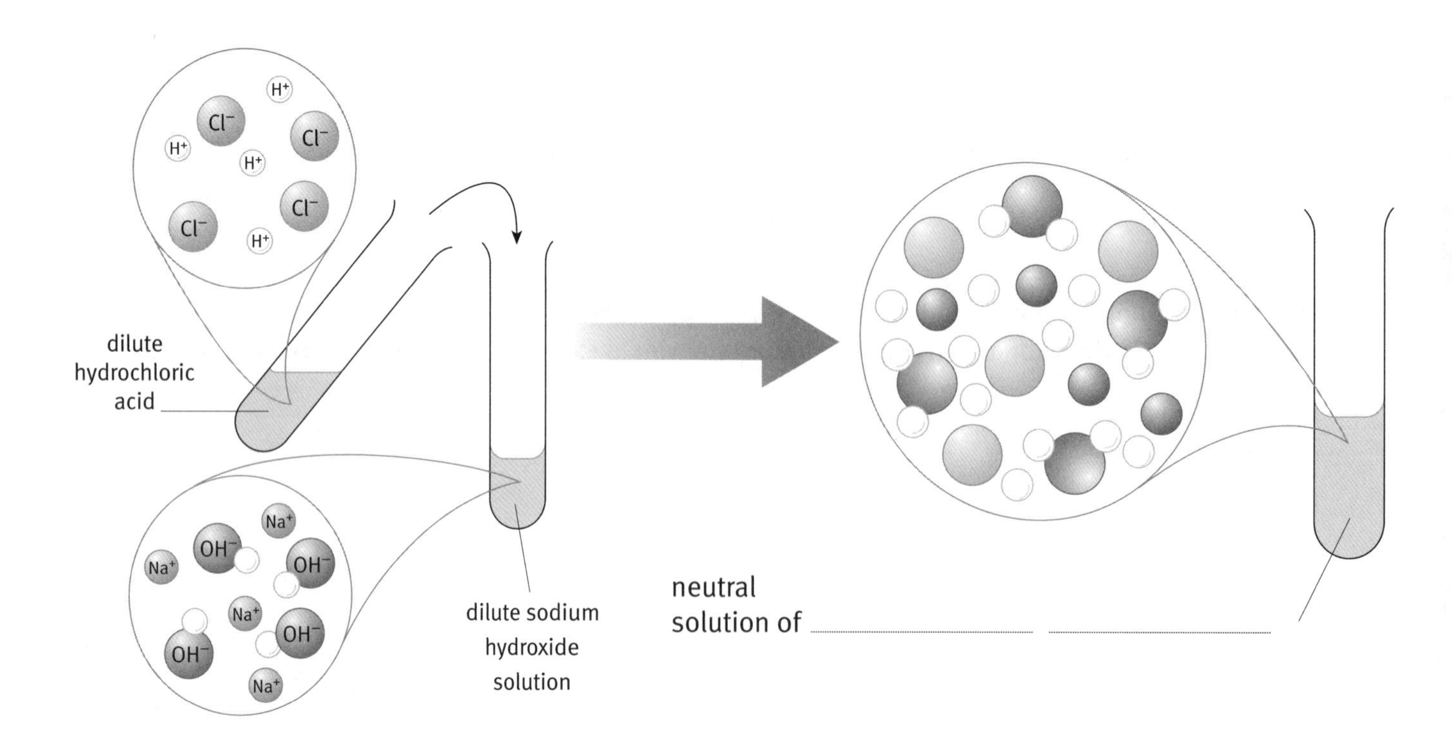

This page is blank

Chemistry for a sustainable world C7

1 Methane molecules

Complete this table, which shows three ways of picturing a molecule of methane. Add these terms in the correct places as column headings:

- ball-and-stick model
- molecular formula
- structural formula

CH_4		

2 Alkane formulae

Complete this table by adding the missing information about three alkanes.

Name	**Molecular formula**	**Structural formula**
	C_2H_6	
		H H H | | | H—C—C—C—H | | | H H H
butane		

3 Burning alkanes

Many fuels contain alkanes. Alkanes burn in air. They react with oxygen. Write a balanced equation, with state symbols, for the reaction of ethane burning in plenty of air.

....

4 Alkane properties

Use the terms in the box to complete the summary of the properties of alkanes.

alkane	crude oil	gases	hydrocarbons	oily	single bonds	water

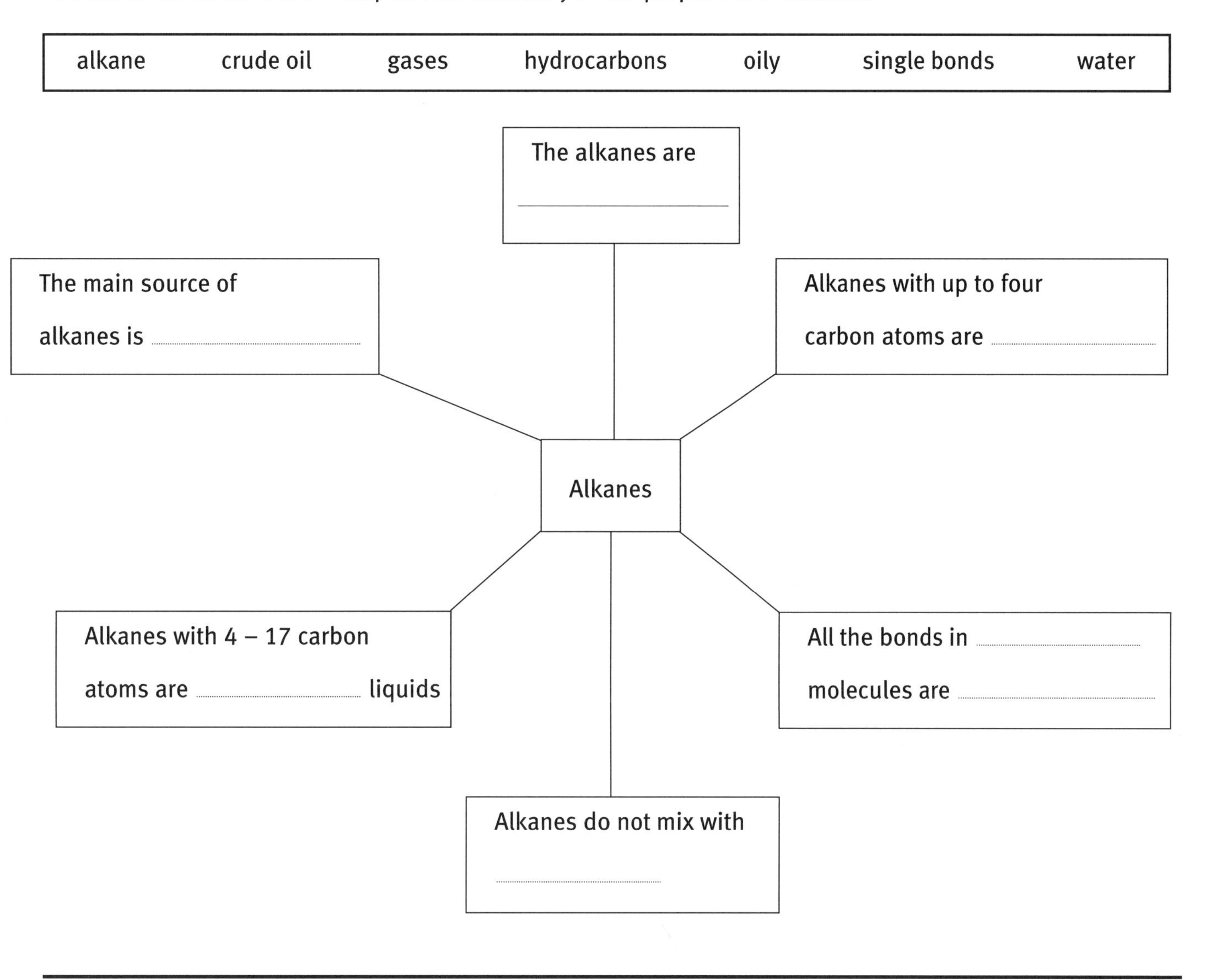
The alkanes are ……
The main source of alkanes is ……
Alkanes with up to four carbon atoms are ……
Alkanes
Alkanes with 4 – 17 carbon atoms are …… liquids
All the bonds in …… molecules are ……
Alkanes do not mix with ……

5 Alkane structure and reactivity

Add notes and labels to this diagram to explain why alkanes do not react with common laboratory reagents, such as acids and alkalis.

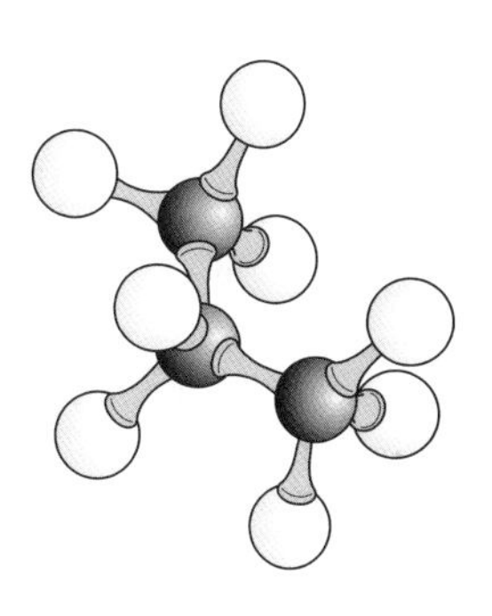

6 Methanol and ethanol

Complete this table. In the models, colour the carbon atoms black and the oxygen atoms red.

	Methanol	Ethanol
Molecular formula		
Ball-and-stick model		
Examples of uses		

7 Alcohols compared to water and alkanes

Complete this table to compare methanol with water and methane. In the models, colour the carbon atoms black and the oxygen atoms red.

	Water	Methanol	Methane
Molecular formula			
Ball-and-stick model			
State at room temperature			
Boiling point			−161 °C
Ease of mixing or dissolving in water			
Explanation in terms of the strength of the attraction between molecules			

8 Functional groups

In this diagram of an ethanol molecule, colour the carbon atoms black and the oxygen atoms red. Label and annotate the diagram to show:

- the functional group
- the bonds which are reactive
- the bonds which are not reactive.

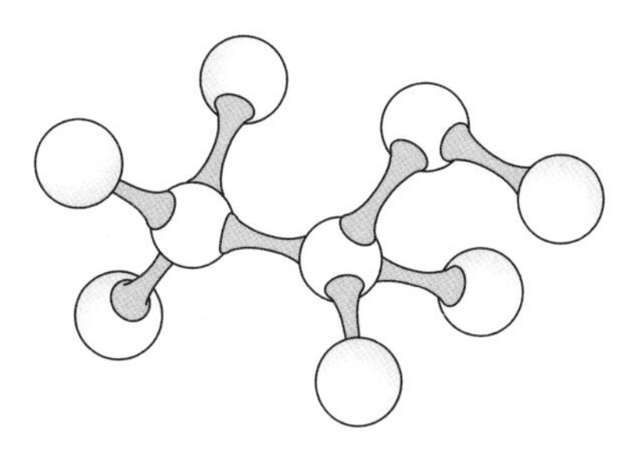

9 Reactions of alcohols

a In some ways alcohols are like water. However, alcohols burn and water does not. Explain why alcohols can burn and state the products of burning if there is plenty of air.

....

....

....

....

b Write a balanced equation for the reaction of ethanol burning in air.

....

c Complete this table to compare the reactions of three chemicals with sodium.

	Water	**Methanol**	**Methane**
Does the chemical react with sodium? If yes, what are the products?			
Formula of the product that contains sodium (if there is one)			

10 Natural occurrence of organic acids

Draw lines to link the names of carboxylic acids to where they can be found naturally.

Acetic acid (ethanoic acid)	In the fats of goats milk and in stale sweat
Butyric acid (butanoic acid)	Orange and lemon juice
Caproic acid	Rancid butter and vomit
Citric acid	Sour milk
Lactic acid	Vinegar

---

11 Reactions of carboxylic acids

Complete these word equations to show the typical reactions of organic acids with metals, metal oxides, and hydroxides, and metal carbonates.

methanoic acid + ... → magnesium methanoate + hydrogen

methanoic acid + sodium hydroxide → .... + ...

ethanoic acid + ... → copper ethanoate + water

ethanoic acid + potassium carbonate → .... + ... + ...

---

12 pH of solutions of acids

Draw lines to match each solution to its pH value. Some solutions listed have the same pH value.

Dilute acetic acid (ethanoic acid)	pH 7
Dilute hydrochloric acid	pH 3
Vinegar	pH 1
Pure water	

13 Carboxylic acid formulae and structures

a Complete this table. In the models, colour the carbon atoms black and the oxygen atoms red.

	Methanoic acid	**Ethanoic acid**
Molecular formula		
Structural formula		
Ball-and-stick model	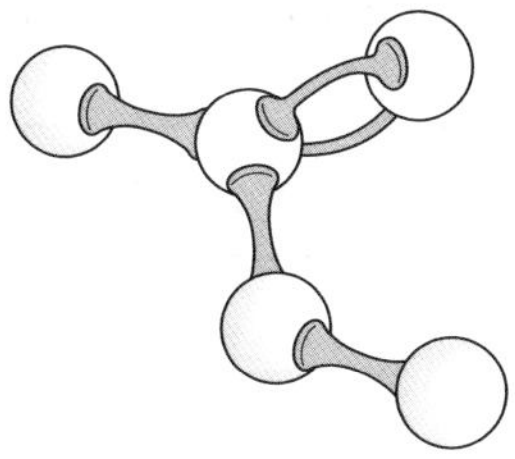	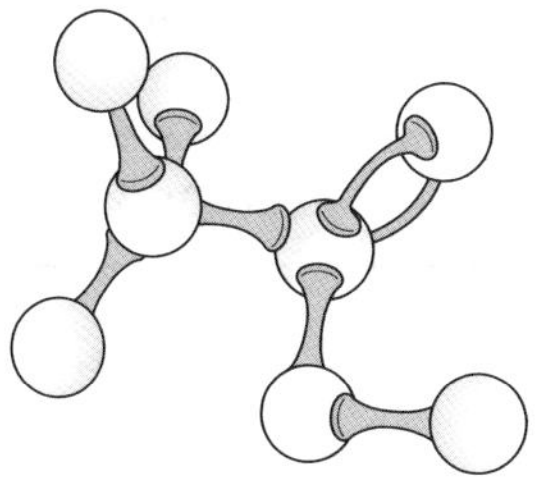

b What is the functional group in an organic acid?

....

....

c i Write a symbol equation to show how ethanoic acid ionizes when it dissolves in water.

ii Which of the products of the ionization of ethanoic acid makes the solution acidic?

....

d Draw a ring round the formulae below that represent carboxylic acids.

CH_3OCH_3 HCOOH $CH_3CH_2CH_2CH_2OH$ $CH_3CH_2CH_2COOH$

14 Esters

a Give one word to describe the typical smell of many simple esters.

....

b Give examples of **three** foods we eat that taste of mixtures of esters.

....

c Give two uses of esters.

- ....
- ....

d Use the names of the chemicals in the box below to write a word equation for the formation of an ester.

water	ethanol	ethyl butanoate	butanoic acid

....

e Complete this word equation:

pentanol + ethanoic acid → .... + ...

f Label the diagrams below to show how to make a small sample of an ester and then smell the product.

The words in the box will help you.

acid	alcohol	catalyst	carboxylic acid	hot water
neutralize leftover acid	reaction of mixture after warming	sodium carbonate solution		

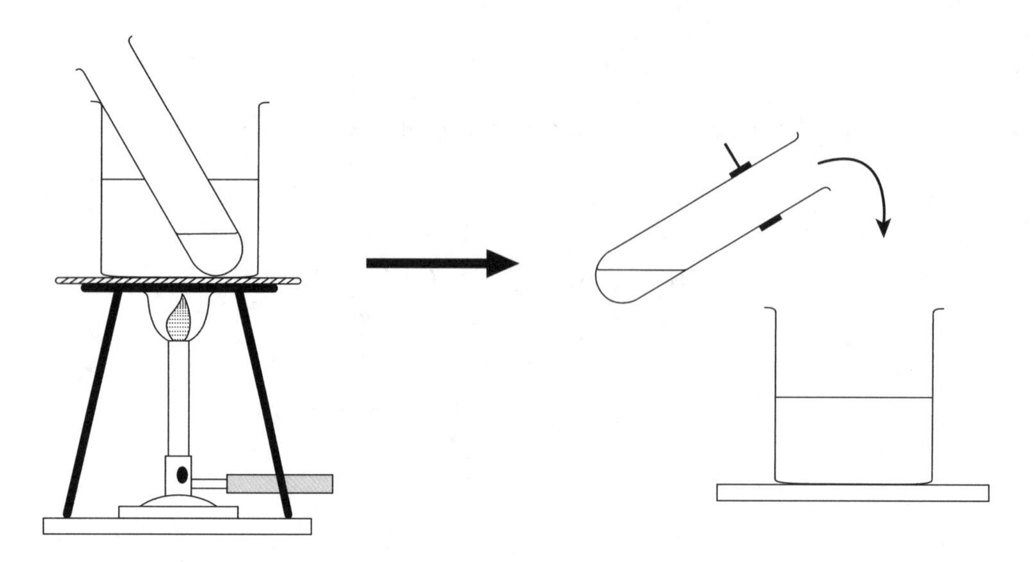

15 Preparation of an ester

This flow diagram shows the procedure used to make a sample of an ester on a laboratory scale.

Annotate (label) the diagram. The words in the box below will help you. You may choose to use some words more than once.

aqueous reagent to remove impurities	catalyst	drying agent	ethanoic acid
ethanol	ester from tap funnel	heat	impure ester layer
concentrated sulfuric acid	distillation flask	impure product	pure ethyl ethanoate
reflux condenser	tap funnel	thermometer	water condenser

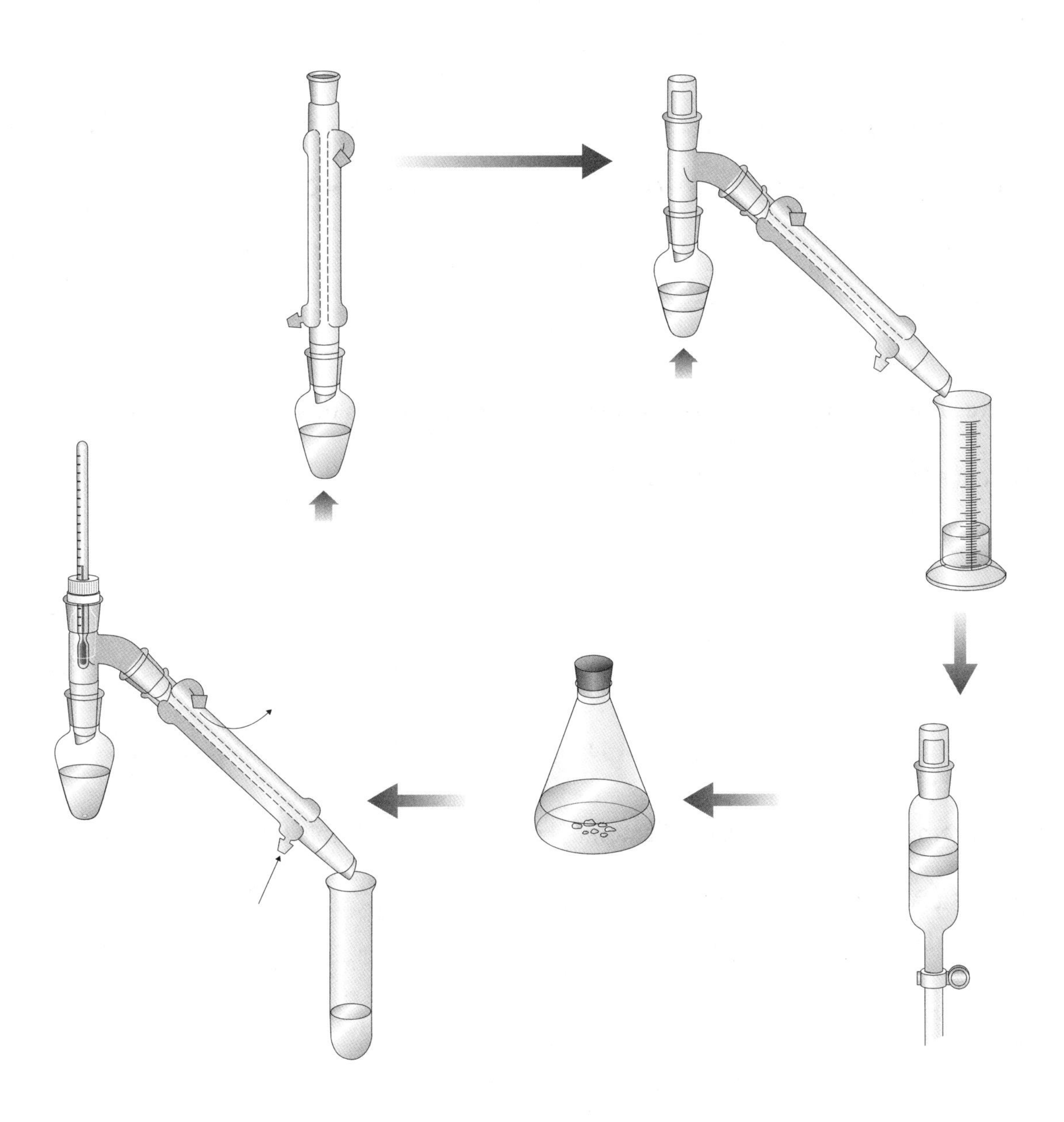

16 Fats and oils

a Why are fats and oils important to plants and animals?

....

b These diagrams show the structures of a molecule from a fat and a molecule from an oil. Label both diagrams to show: an ester link, a hydrocarbon chain, and the part of the molecule which comes from glycerol.

$$\begin{array}{l}
\quad H \qquad\quad O \\
\quad | \qquad\quad \| \\
H-C-O-C-CH_2-CH_2-CH_2-CH_2-CH_2-CH_2-CH_2-CH_2-CH_2-CH_2-CH_2-CH_2-CH_3 \\
\quad | \qquad\quad O \\
\quad | \qquad\quad \| \\
H-C-O-C-CH_2-CH_2-CH_2-CH_2-CH_2-CH_2-CH_2-CH_2-CH_2-CH_2-CH_2-CH_2-CH_3 \\
\quad | \qquad\quad O \\
\quad | \qquad\quad \| \\
H-C-O-C-CH_2-CH_2-CH_2-CH_2-CH_2-CH_2-CH_2-CH_2-CH_2-CH_2-CH_2-CH_2-CH_3 \\
\quad | \\
\quad H
\end{array}$$

Diagram A

$$\begin{array}{l}
\quad H \qquad\quad O \\
\quad | \qquad\quad \| \\
H-C-O-C-CH_2-CH_2-CH_2-CH_2-CH_2-CH_2-CH_2-CH=CH-CH_2-CH=CH-CH_2-CH=CH-CH_2-CH_3 \\
\quad | \qquad\quad O \\
\quad | \qquad\quad \| \\
H-C-O-C-CH_2-CH_2-CH_2-CH_2-CH_2-CH_2-CH_2-CH=CH-CH_2-CH=CH-CH_2-CH=CH-CH_2-CH_3 \\
\quad | \qquad\quad O \\
\quad | \qquad\quad \| \\
H-C-O-C-CH_2-CH_2-CH_2-CH_2-CH_2-CH_2-CH_2-CH=CH-CH_2-CH=CH-CH_2-CH=CH-CH_2-CH_3 \\
\quad | \\
\quad H
\end{array}$$

Diagram B

c Are the hydrocarbon chains in molecule A saturated or unsaturated? Is this the structure of a molecule from a fat or an oil? Give your reasons. ....

....

....

....

....

d Are the hydrocarbon chains in molecule B saturated or unsaturated? Is this the structure of a molecule from a fat or an oil? Give your reasons. ....

....

....

....

....

17 Exothermic reactions

a The reaction of magnesium with dilute hydrochloric acid is an exothermic reaction. Use the words in the box to label the diagram below, and explain what is meant by the term 'exothermic reaction'.

dilute hydrochloric acid	energy given out	exothermic	magnesium
magnesium chloride solution			

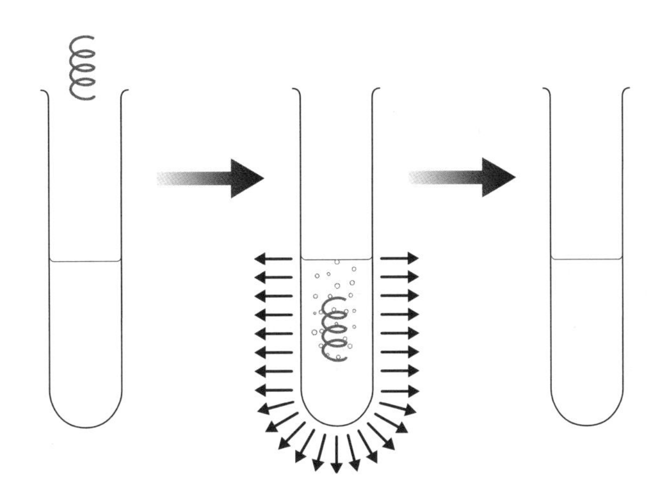

b Write a balanced symbol equation, with state symbols, for the reaction of magnesium with dilute hydrochloric acid.

....

c Use parts of the symbol equation to label the energy level diagram below by adding the reactants and products to the diagram. Complete the labelling of the diagram.

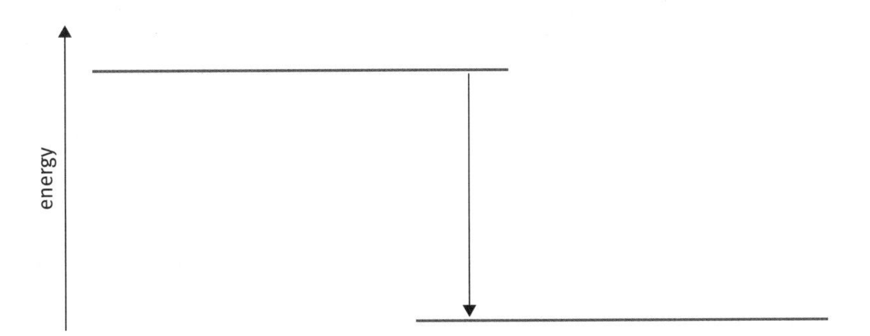

d Give two more examples of exothermic changes:

- ....
- ....

18 Endothermic reactions

a The reaction of citric acid with sodium hydrogencarbonate is an endothermic reaction. Use the words in the box to label the diagram below, and explain what is meant by the term 'endothermic reaction'.

citric acid	endothermic	energy taken in	sodium citrate solution
sodium hydrogencarbonate			

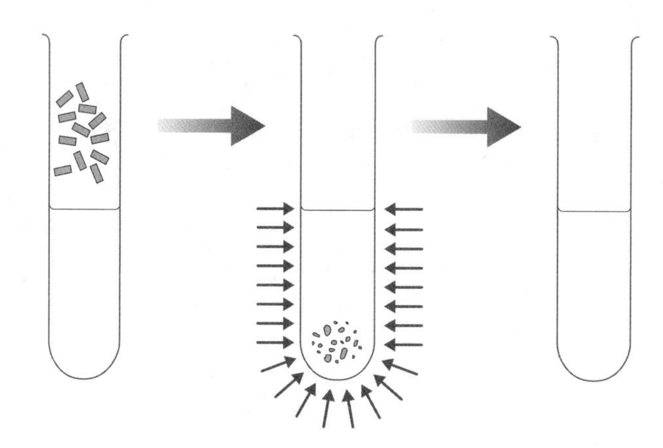

b Write a word equation, with state symbols, for the reaction of citric acid with sodium hydrogencarbonate.

....

c Use parts of the word equation to label the energy level diagram below by adding the reactants and products to the diagram. Complete the labelling of the diagram.

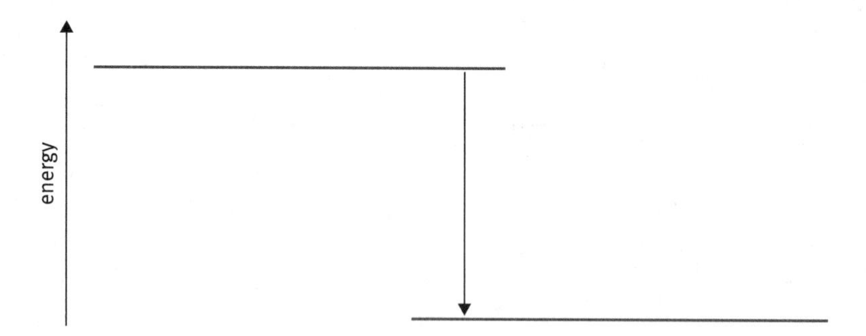

d Give two more examples of endothermic changes:

- ....
- ....

19 Bond breaking and bond forming

a Use the words in the box to label the diagram below. Colour the oxygen atoms red.

bond broken during reaction	bond formed during reaction	hydrogen molecule
oxygen molecule	water molecule	

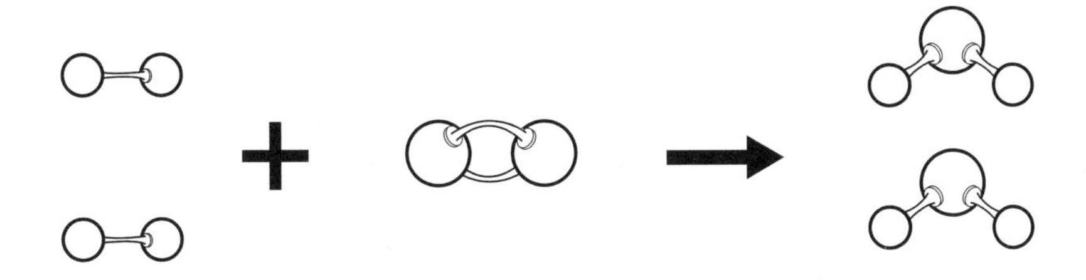

b Explain why the reaction of hydrogen with oxygen is an exothermic reaction.

....

....

c Use the data in the table to complete the labelling of the diagram below.

Process	**Energy change for breaking all the bonds in the formula mass of the chemical**
Breaking all the H—H bonds in hydrogen	434 kJ needed
Breaking all the Br—Br bonds in bromine	193 kJ needed
Breaking all the H—Br bonds in hydrogen bromide	366 kJ needed

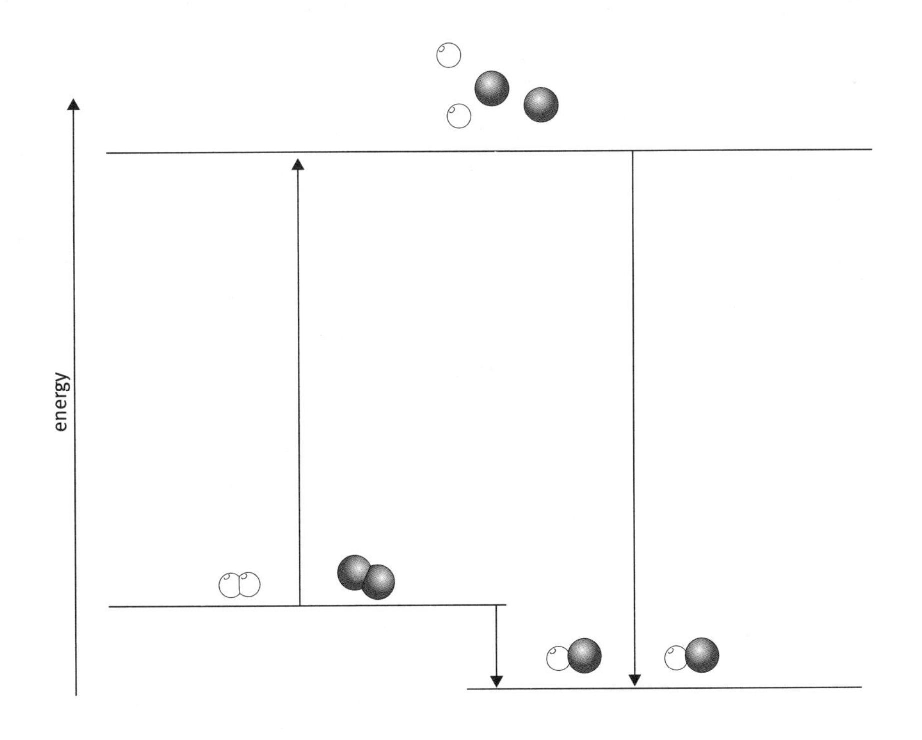

20 Activation energies

a With the help of some or all of the words and phrases in the table, write a paragraph to explain what is meant by the term activation energy.

new bonds form	high energy	bonds break	unsuccessful collision
reactant molecules	millions of collisions per second	low energy	collisions
not all collisions lead to reaction	minimum energy	successful collision	product molecules

b Use the idea of activation energy to explain why reactions go faster as the temperature rises.

21 Reversible changes

a For each of these pairs of equations, indicate the conditions needed to make the change go in the direction shown in the equation.

Conditions needed to make the change happen in the direction shown:

$H_2O(l) \rightarrow H_2O(g)$ ....

$H_2O(g) \rightarrow H_2O(l)$ ....

$CuSO_4.5H_2O(s) \rightarrow CuSO_4(s) + 5H_2O(l)$ ....

$CuSO_4(s) + 5H_2O(l) \rightarrow CuSO_4.5H_2O(s)$ ....

$NH_3(g) + HCl(g) \rightarrow NH_4Cl(s)$ ....

$NH_4Cl(s) \rightarrow NH_3(g) + HCl(g)$ ....

b Hot iron reacts with steam to form the iron oxide, Fe_3O_4 and hydrogen.

Write a balanced symbol equation for the reaction.

....

This reaction is reversible. Label the diagram to show how to demonstrate the reverse reaction.

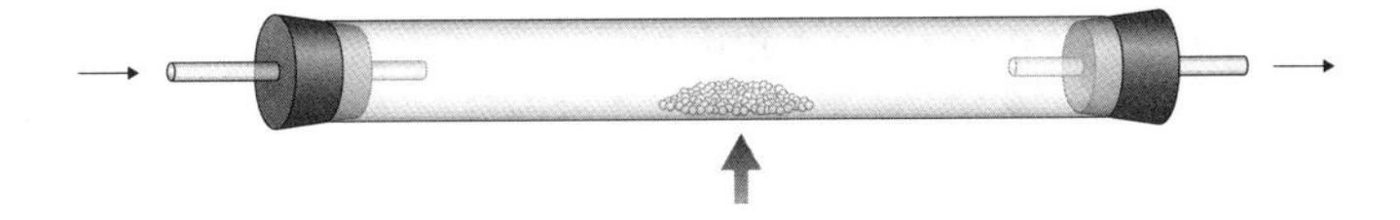

Write a balanced symbol equation for the reverse reaction.

....

22 Dynamic equilibrium

a The diagram below shows a crystal of iodine dissolving in hexane. The solution formed is then shaken with a solution of potassium iodide in water. Colour the solutions.

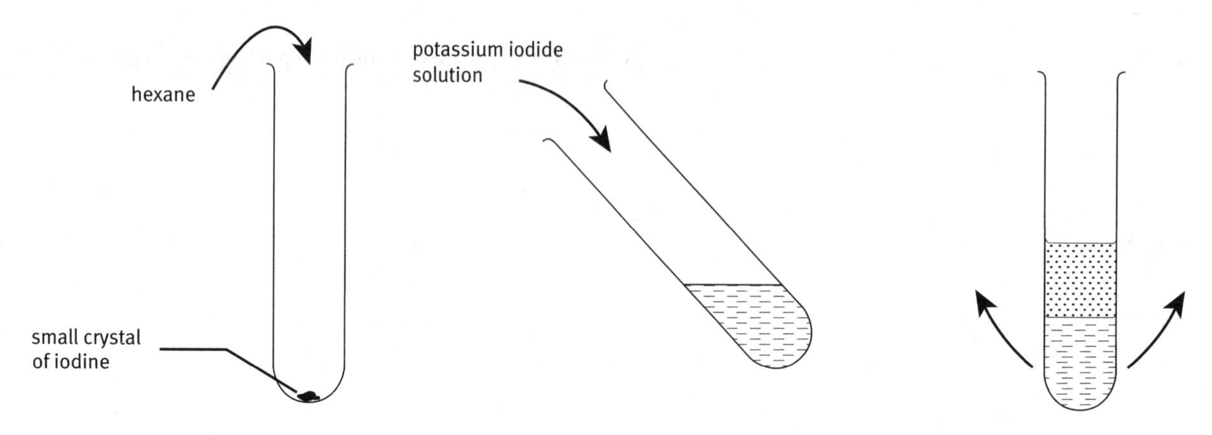

Explain why there is some iodine in both layers, however long the two solutions are shaken up together. ....

....

....

b The diagram below shows a crystal of iodine dissolving in aqueous potassium iodide. The solution formed is then shaken with hexane. Colour the solutions.

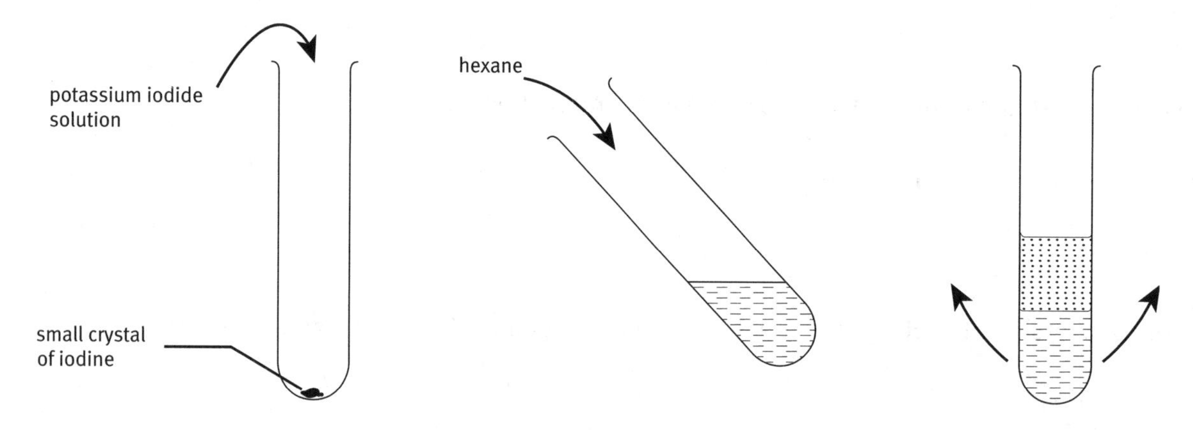

Explain why the two layers look the same after shaking as they did when the crystal was first dissolved in hexane and then shaken with potassium iodide solution. ....

....

....

c Why is the term 'dynamic equilibrium' used to describe the state reached when solutions of iodine in hexane and in aqueous potassium iodide are shaken up together? ....

....

....

23 Dynamic equilibrium on a molecular scale

The diagrams below show what happens to iodine molecules on shaking a solution of iodine in hexane (violet) with a solution of potassium iodide in water (which starts colourless). Colour the diagrams. Under each diagram explain what is happening to the molecules.

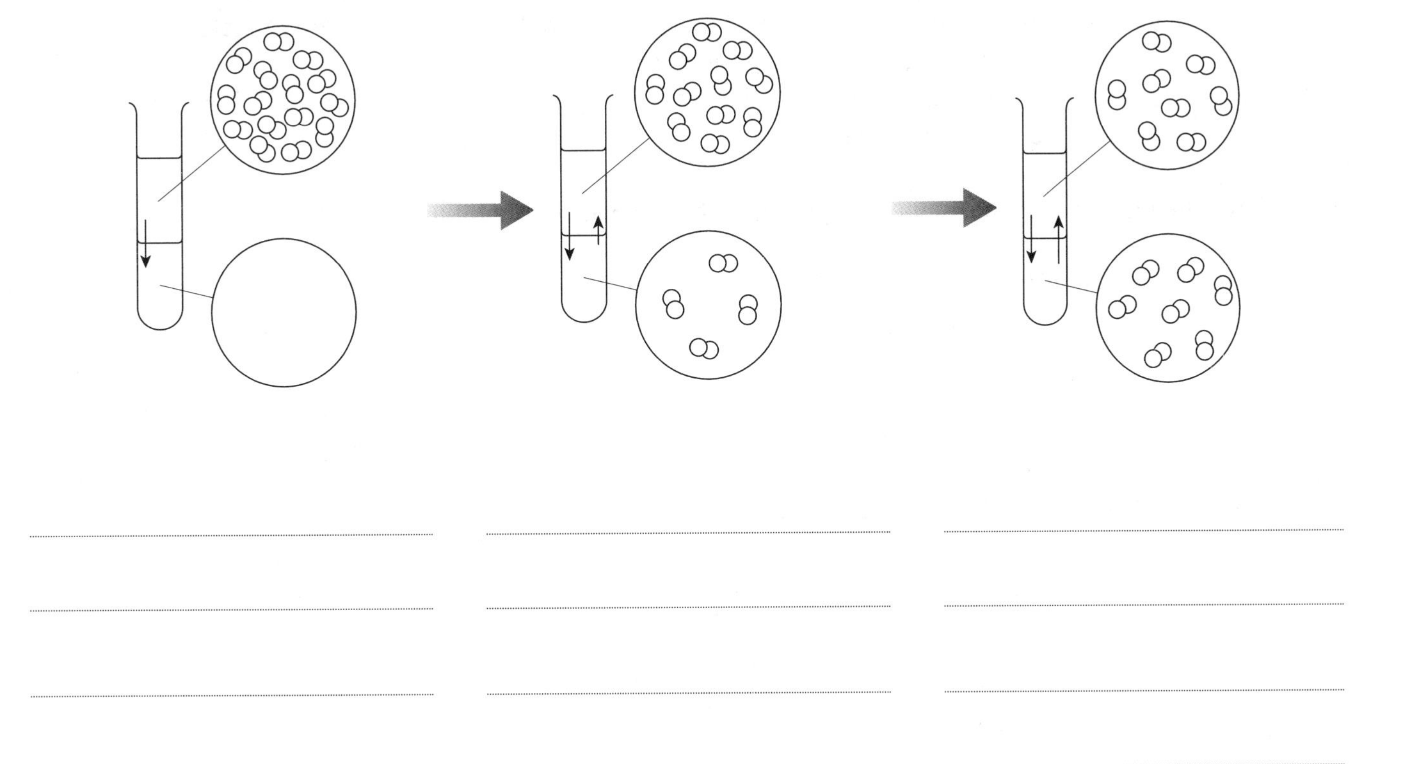

24 Strong and weak acids

Fill in the blanks to complete these sentences and equations.

- Hydrogen chloride gas dissolves in water to give a solution of .... acid.

  HCl(g) + water → .... (aq) + Cl^-(aq)

- All of the molecules of hydrogen chloride ionize when the gas dissolves in ....

  Hydrogen chloride is a .... acid. Another example of a strong acid is

  .... acid.

- Carboxylic acids are .... acids. In a dilute solution of ethanoic acid, only about one molecule in a hundred ionizes. In solution there is a dynamic ....

  $CH_3COOH(aq) + H_2O(l) \rightleftharpoons CH_3COO^-(aq) +$ .... (aq)

25 Sampling for analysis

a When taking samples for analysis, it is important to make sure that they are typical of the whole bulk of the material analysed. Analysts have to decide:

- how many samples to collect, and how much of each to collect, to be sure that the samples are representative
- how many times to repeat an analysis on a sample to be sure that the results are reliable
- where, when and how to collect samples of the material
- how to store samples and take them to the laboratory to prevent samples 'going off', becoming contaminated, or being tampered with.

Write down the main issues for the person planning the analysis of the following materials.

i Soil in a farmer's field

ii Water in a pond

iii Polluted air in a city street

iv Urine from an athlete

b Give reasons why it is important to have standard procedures for collecting, storing and analysing samples.

26 Paper and thin-layer chromatography

a Label this diagram. It may help to use colours.

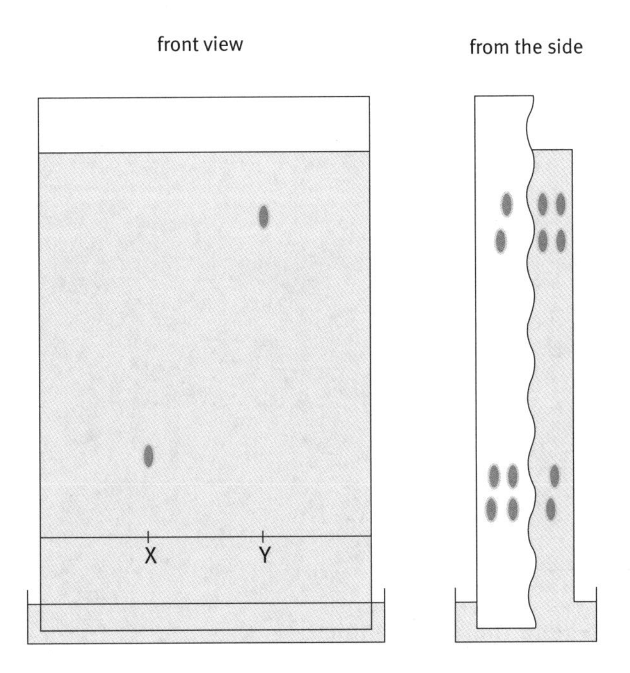

- Label the mobile phase, stationary phase and solvent front.
- Add an arrow to show the direction in which the solvent front moves.
- Add arrows to show that for both chemicals in samples X and Y there is a dynamic equilibrium between the stationary phase and the mobile phase.

i What would happen to a spot of substance on the start line that is not at all soluble in the mobile phase?

....

ii Explain why sample Y moves further than sample X.

....

....

b Complete this table to compare paper and thin-layer chromatography

	Paper chromatography	**Thin-layer chromatography**
Stationary phase		
Mobile phase		
Speed of separation		
Quality of separation		
Qualitative or quantitative?		

27 Interpreting chromatograms

The diagram below is a chromatogram of an extract from a supermarket curry sauce (S). Four reference samples of permitted colours have also been run on the chromatogram (E102, E110, E122 and E124).

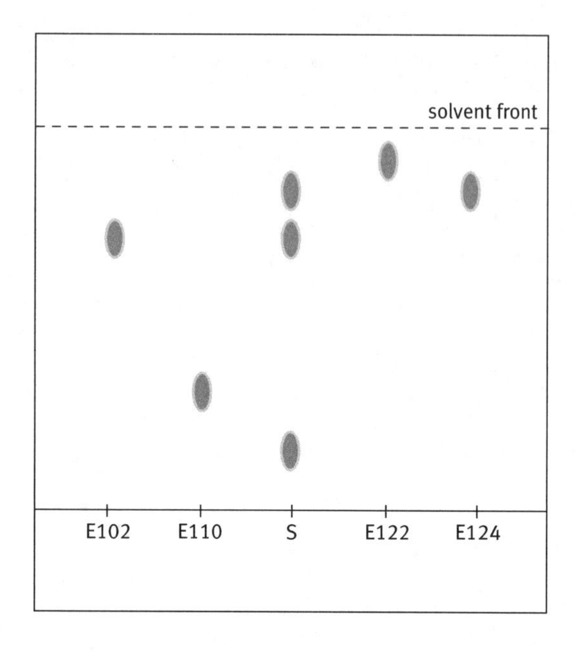

a How many coloured chemicals were there in sample S? ....

b Which permitted additives were present in the curry sauce? ....

c Calculate the R_f value for the spot that does not match any of the reference colours.

$$R_f = \frac{\text{distance moved by spot}}{\text{distance moved by solvent front}}$$

....

....

d Under the same conditions, the $R_f = 0.15$ for a banned colouring, Sudan 1. What does the chromatogram show about the colourings in the curry sauce?

....

....

e Suggest two ways which could be used to detect colourless additives on the chromatogram.

- ....
- ....

28 Gas chromatography

a Use the words in the box to label the diagram below, which describes gas chromatography (GC).

chromatogram	column packed with stationary phase	cylinder of carrier gas	detector
flow meter	mobile phase	oven	recorder
sample injected here	vent		

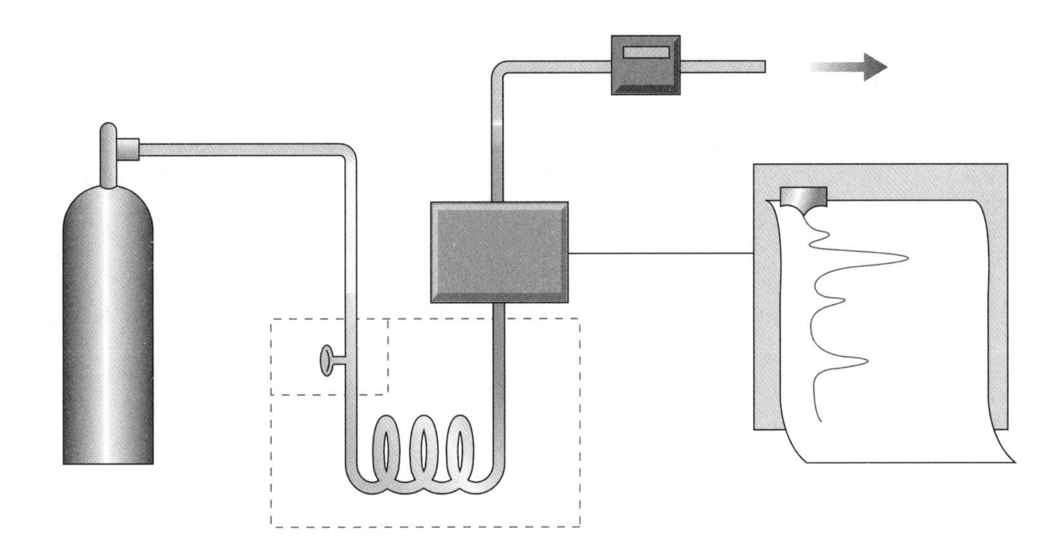

b The diagram below shows a gas chromatogram.

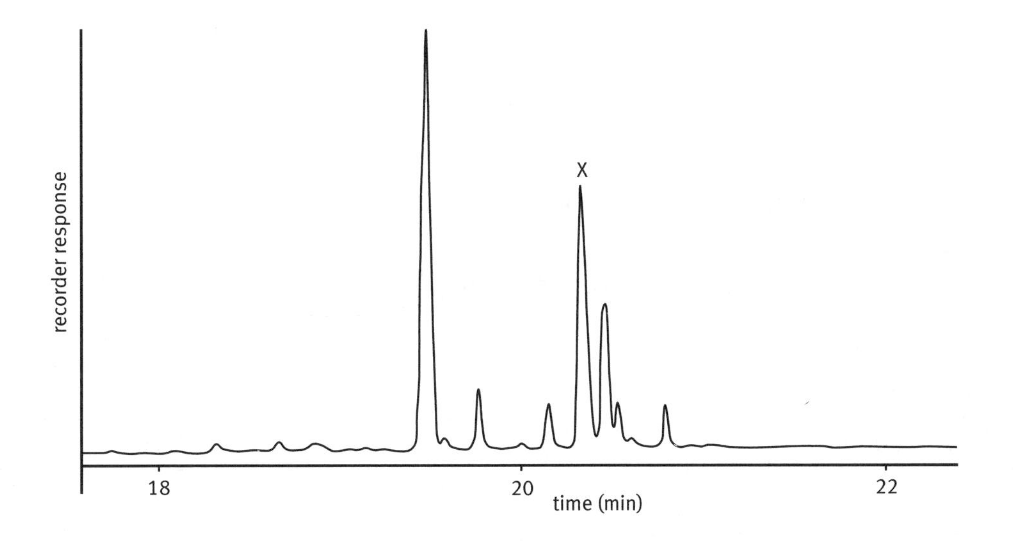

How many chemicals were there in the mixture injected into the GC instrument? ....

Which chemical was present in the largest amounts in the sample? ....

Estimate the retention time of chemical X. ....

....

29 Concentrations of solutions

a Label this sequence of diagrams to show how you would prepare a solution of sodium carbonate, Na_2CO_3, with a concentration of 10.6 g/dm³, assuming that the volume of the graduated flask is 250 cm³.

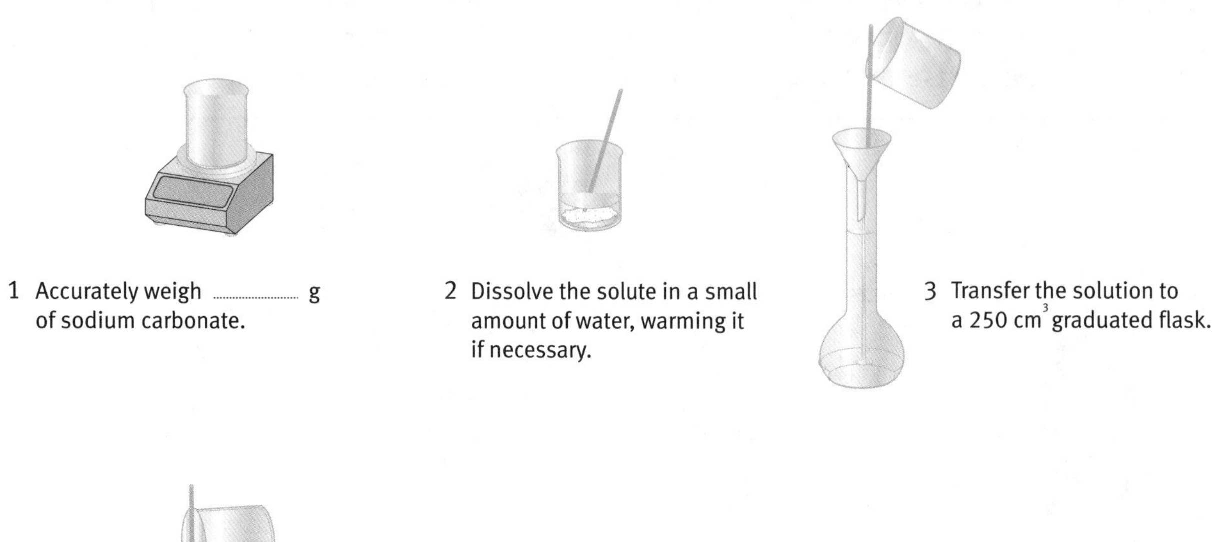

1 Accurately weigh .... g of sodium carbonate.

2 Dissolve the solute in a small amount of water, warming it if necessary.

3 Transfer the solution to a 250 cm³ graduated flask.

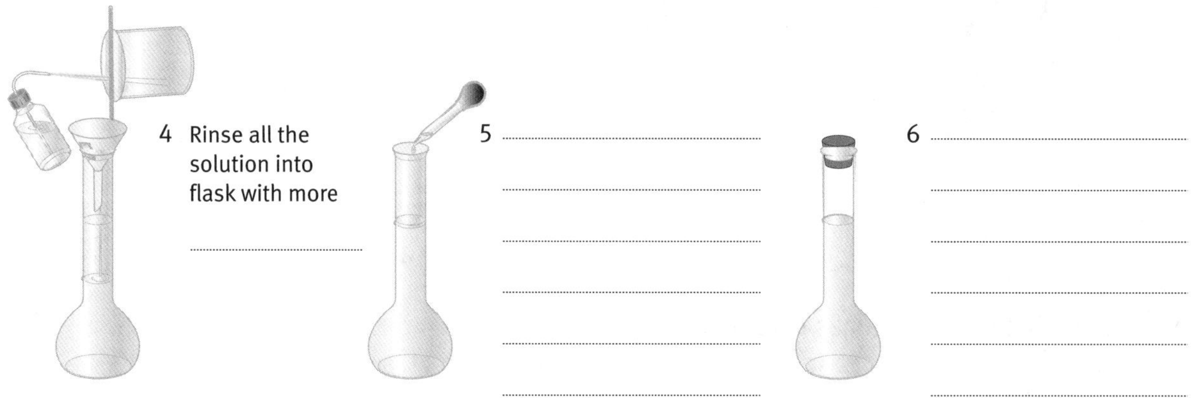

4 Rinse all the solution into flask with more ....

5 ....
....
....
....
....
....

6 ....
....
....
....
....
....

b Complete the table to show the concentrations of these solutions in g/dm³.

Solution	**Concentration in g/dm³**
20.0 g magnesium sulfate in 500 cm³ solution	
4.5 g potassium hydroxide in 250 cm³ solution	
0.5 g sodium sulfate in 10 cm³ solution	
1.25 g silver nitrate in 50 cm³ solution	

c Complete the table to show the mass of solute in these volumes of solutions.

Sample of solution	**Mass of dissolved solute in g**
100 cm³ of a 25.0 g/dm³ zinc sulfate solution	
50 cm³ of a 10.0 g/dm³ lead nitrate solution	
10 cm³ of a 22.5 g/dm³ magnesium chloride solution	
2.5 cm³ of a 16.0 g/dm³ barium nitrate solution	

30 Use of a pipette

A pipette is only accurate if it is used correctly. Suggest reasons for the following questions. Each question is designed to remind analysts about correct checks and procedures.

a Have you rinsed the pipette with the solution you are going to measure out?

b Have you made sure that there are no air bubbles in the narrower parts of the pipette?

c Have you wiped the outside of the pipette to remove solution on the outside of the glass before running out the liquid?

d Have you lined up the meniscus with the graduation mark correctly?

---

31 Use of a burette

A burette is only accurate if it is used correctly. Suggest reasons for the following questions. Each question is designed to remind analysts about correct checks and procedures.

a Have you checked that the burette is clean before you start?

b Have you rinsed the burette with the solution you are going to measure before filling it?

c Have you read the burette correctly and taken both readings?

d Have you left a drop hanging from the tip of the burette after running the solution into the flask?

32 A titration to analyse vinegar

a The diagrams below show steps in a titration to measure the concentration of acetic acid (ethanoic acid) in vinegar. Use the words in the box to label the diagrams. You may use the words more than once.

conical flask	burette	indicator	pipette	sodium hydroxide	vinegar

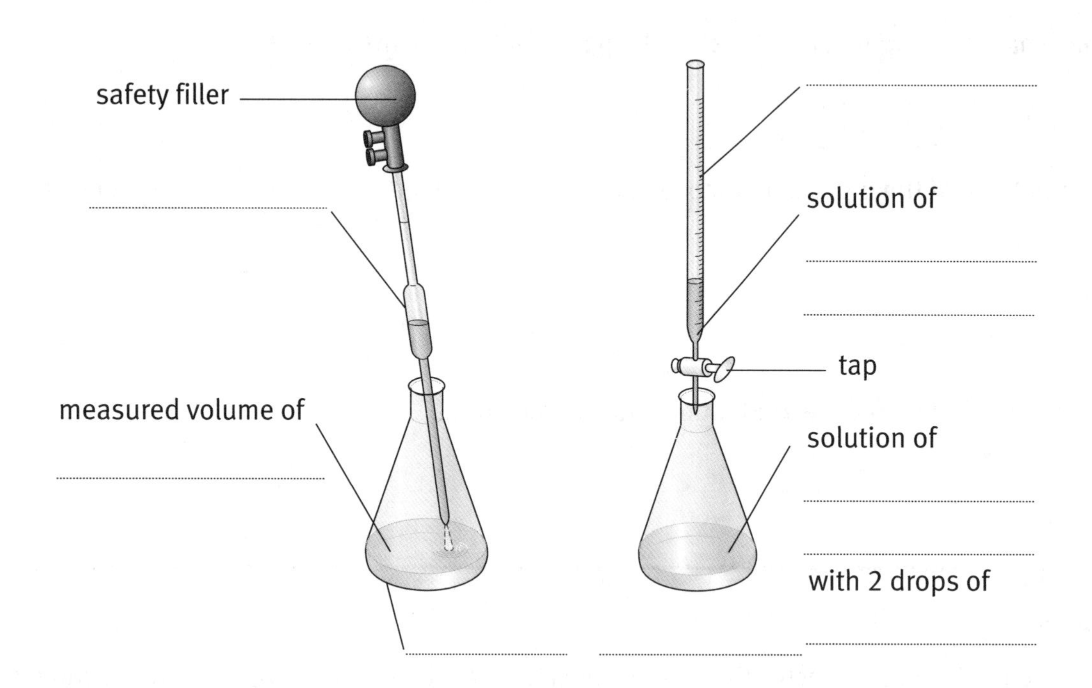

b The table below shows the results of a series of titrations to measure the concentration of acetic acid in vinegar. The flask contained 10.0 cm^3 vinegar and 3 drops of phenolphthalein indicator. The concentration of the sodium hydroxide solution in the burette was 20.0 g/dm^3.

	Rough titration	Titration 1	Titration 2	Titration 3
Second burette reading/cm^3 **First burette reading/cm^3**	17.5 0.0	22.00 5.00	19.00 2.10	20.10 3.20
Volume of NaOH(aq) added/cm^3				

i Complete the bottom row of the table.

ii Draw a ring round the values you use to work out an average value.

iii The average value for the volume of alkali added = ... cm^3

iv Use this formula to work out the concentration of the vinegar.

$$\text{Vinegar concentration (g/dm}^3) = \frac{3}{2} \times \text{NaOH concentration (g/dm}^3) \times \frac{\text{volume of NaOH (cm}^3)}{\text{volume of vinegar (cm}^3)}$$

Concentration of vinegar = ...

v The percentage of acetic acid by mass in the vinegar = ...

33 Interpreting titration results

An analyst carried out a titration to find the concentration of limewater. Limewater is a saturated solution of calcium hydroxide, $Ca(OH)_2$ in water. The analyst measured out 20.0 cm^3 samples of limewater and then carried out titrations with dilute hydrochloric acid. The concentration of the acid was 1.46 g/dm^3 HCl(aq). The average titre was 25.0 cm^3 of the dilute hydrochloric acid. Follow these steps to work out the concentration of calcium hydroxide in limewater.

a Write the balanced equation for the reaction which takes place during the titration.

....

b Work out the relative formula masses of calcium hydroxide and hydrochloric acid

(Relative atomic masses: Ca = 40, O = 16, H = 1, Cl = 35.5).

$Ca(OH)_2$ ....

HCl ....

c Calculate the mass of HCl in the 25.0 cm^3 of the dilute hydrochloric acid added from the burette.

....

....

d Use the equation and the reacting masses to calculate the mass of calcium hydroxide that reacts with the HCl added from the burette.

....

....

e This is the mass of calcium hydroxide in 20.0 cm^3 of limewater. Calculate the concentration of calcium hydroxide in limewater in g/dm^3.

....

....

34 Accurate quantitative analysis

Below are stages in a quantitative analysis. For each stage, show how this applies to a titration or explain why it is important.

- Measuring out accurately a specified mass or volume of the sample

- Working with replicate samples

- Dissolving samples quantitatively

- Measuring a property of the solution quantitatively

- Calculating a value from the measurements

- Estimating the degree of uncertainty in the results

35 Sources of raw materials

Inorganic compounds are obtained from never-lived sources. Organic compounds are obtained from living organisms and non-living sources. Sort the raw materials into the table.

air	bauxite	coal	crude oil	haematite	limestone	natural gas
phosphate rock	quartz	starch	rock salt	sugar cane	sunflower oil	
water	wood					

Never-lived sources	Once-lived sources	Living sources

36 Bulk and fine chemicals

a Fill in the missing names and formulae in the table of bulk chemicals.

Bulk chemicals	
Formula	**Name**
NH_3	ammonia
H_2SO_4	
	sodium hydroxide
H_3PO_4	phosphoric acid
C_2H_4	ethene

b Complete the table using these examples of fine chemicals.

carotene	citral	ibuprofen	glyphosate

Fine chemicals	
Type	**Example**
medical drugs	
agrochemicals	
food additives	
fragrances	

37 Regulation of the chemical industry

The UK government regulates chemical companies. For each of these areas, suggest a reason why it is important to have rules to protect the public, people at work, or the environment.

- Choice of raw materials for manufacturing ....

....

- Transporting chemicals ....

....

- Storing chemicals ....

....

- Getting rid of chemical waste ....

....

- Labelling of packs of chemicals ....

....

---

38 Chemical plants

Use the words in the box to complete the labelling of the diagram below, which summarizes key aspects of a typical chemical process in industry.

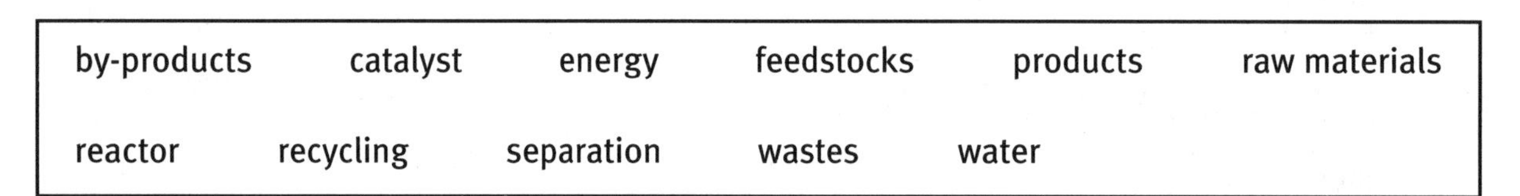

by-products
catalyst
energy
feedstocks
products
raw materials
reactor
recycling
separation
wastes
water

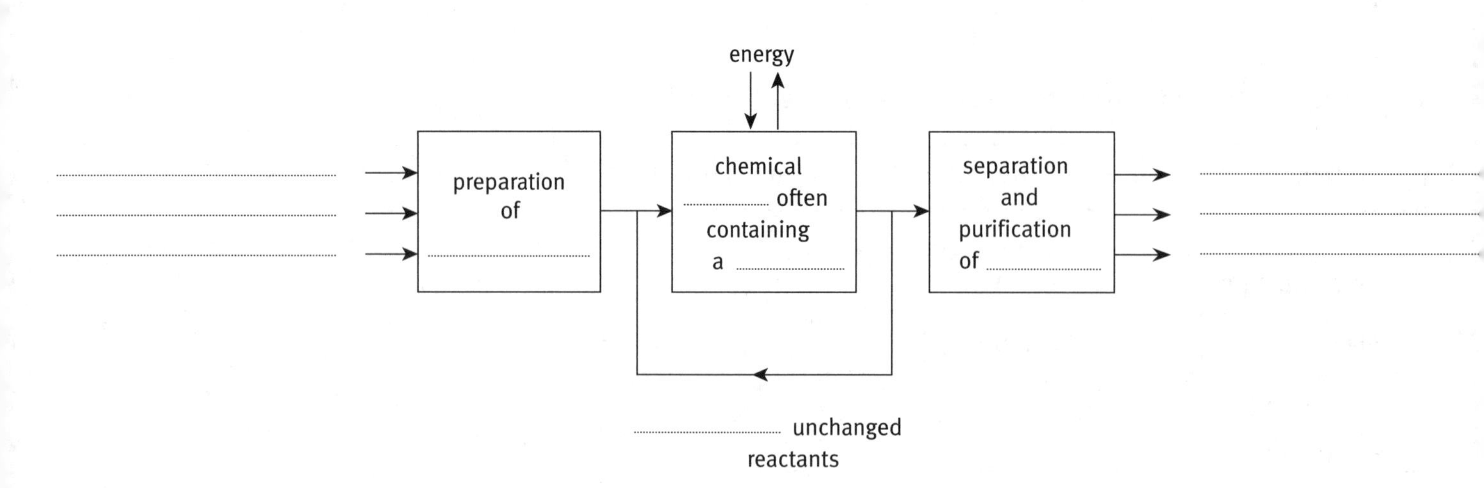

39 Types of feedstock

Use two colours for the key and then shade the areas in the grid with the colours to show which of these chemical feedstocks are renewable and which are not.

Key: ☐ Renewable ☐ Not renewable

ethene from refining crude oil fractions	methanol made from natural gas and steam	sulfur from the purification of natural gas
succinic acid from the fermentation of wastes from papermaking	sodium chloride (brine) from the dissolving of underground salt deposits	ethanol from the fermentation of sugars
oxygen from the air	lactic acid from the fermentation of beet sugar	naptha from distilling crude oil

40 Sustainable chemistry

Give examples from the chemical industry to show that the following can contribute to making chemical processes more sustainable.

- Avoiding chemicals that are hazardous to health ....

....

- Managing energy inputs and outputs ....

....

- Recycling chemicals ....

....

- Finding new uses for by-products ....

....

- Reducing waste products ....

....

41 Atom economy

There are two processes for making the chemical ethylene oxide. Complete the tables for the two processes and calculate the atom economies.

(Relative atomic masses: H = 1, C = 12, O = 16, Cl = 35.5, Ca = 40)

Method 1: The two-step route

$$C_2H_4 + Cl_2 + H_2O \rightarrow CH_2ClCH_2OH + HCl$$

$$CH_2ClCH_2OH + HCl + Ca(OH)_2 \rightarrow \overset{O}{CH_2{-}CH_2} + CaCl_2 + H_2O$$

Formulae of chemicals used	Symbols of atoms that end up in the product	Relative mass of atoms that end up in the product	Symbols of atoms that do not end up in the product	Relative mass of atoms that do not end up in the product
Totals				

Atom economy ....

Method 2: One-step route with a catalyst

$$2C_2H_4 + O_2 \rightarrow 2\overset{O}{CH_2{-}CH_2}$$

Formulae of chemicals used	Symbols of atoms that end up in the product	Relative mass of atoms that end up in the product	Symbols of atoms that do not end up in the product	Relative mass of atoms that do not end up in the product
Totals				

Atom economy ....

42 Catalysts

a Catalysts provide a different route for a reaction with a lower activation energy. Why does this make the reaction go faster?

b Industrial catalysts speed up reactions. Why does this mean that a more efficient catalyst can allow chemical plants to operate with smaller reactors?

c Good catalysts are highly selective. Why can this help to make a chemical process 'greener'?

d Give two examples of catalysts used in the chemical industry and state what they are used for.

➲ Example 1:

➲ Example 2:

43 Making ethanol from petrochemicals

Producing the feedstock, ethene

a Heating ethane at a high temperature breaks up the molecules and produces ethene, C_2H_4, and hydrogen. Write a balanced equation for the reaction.

....

The sources of ethane are natural gas and the distillation of crude oil.

Making ethanol from ethene

A mixture of ethene and steam under pressure combines to make ethanol in the presence of a phosphoric acid catalyst. About 5% of the mixture is converted to ethanol as the compressed gases pass through the catalyst.

b Complete the equation for the reaction

$$\begin{matrix} H & & H \\ & C = C & \\ H & & H \end{matrix} \quad + \quad \text{..............} \quad \longrightarrow \quad \begin{matrix} & | & & | & \\ — & C & — & C & — \\ & | & & | & \end{matrix}$$

c What is the theoretical atom economy for this reaction? ....

d Complete the labelling of this flow diagram for the process.

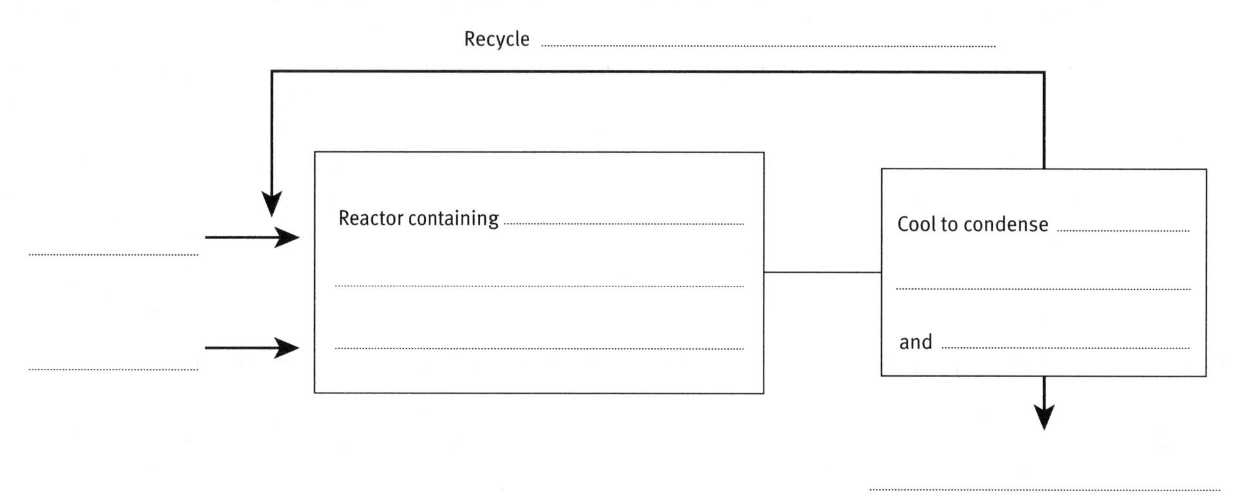

e Why is it necessary to recycle ethene in the process?

....

f Overall, the yield of ethanol from ethene is 95% of the theoretical yield. What mass of ethanol would be produced from 50 tonnes of ethene?

....

....

g Why does the ethanol produced need to be purified?

....

44 Making ethanol by fermentation of sugars

Feedstocks

a Give two examples of raw materials that might be used to make ethanol from biomass:

- a purpose-grown crop ....
- waste material ....

b Use the words in the box to complete the paragraph below.

biomass	catalyse	glucose	long chains	sugars

The chemicals in .... are polymers. They are .... of sugar molecules.

One example, cellulose, is a polymer of the sugar called .... ($C_6H_{12}O_6$). The industry uses water and acid to break down the polymers into simple sugars, such as glucose. After breaking down the polymers, the .... have to be separated from the acid used to .... the reaction.

Fermenting sugars with yeast

Fermentation converts sugars into ethanol and carbon dioxide. Enzymes in yeasts catalyse the reactions. Yeast is a living organism. Fermentation is an example of anaerobic respiration.

c Write a balanced equation for the fermentation of glucose.

....

d Fermentation with yeast works best at temperatures in the range 25 - 37°C. Suggest reasons why fermentation is slow:

- below this temperature range ....
- above this temperature range ....

e Why does fermentation slow down or stop if the alcohol concentration exceeds 14%?

....

f How is it possible to obtain ethanol solutions with a concentration above 14%?

....

45 Making ethanol from biomass with bacteria

Feedstocks

Breaking down biomass with acid can produce a wide range of sugars. These include six-carbon sugars such as glucose ($C_6H_{12}O_6$), and five-carbon sugars such as xylose ($C_5H_{10}O_5$). Yeast is good at converting six-carbon sugars to ethanol but not five-carbon sugars.

a Why is there a need to find new ways to convert sugars to alcohol?

- an economic reason ....

....

- an environmental reason ....

....

Fermenting sugars with GM bacteria

Scientists have used genetic modification to create a bacterium that can convert five-carbon sugars to ethanol. The diagram below shows a process which makes ethanol from biomass using this GM bacterium.

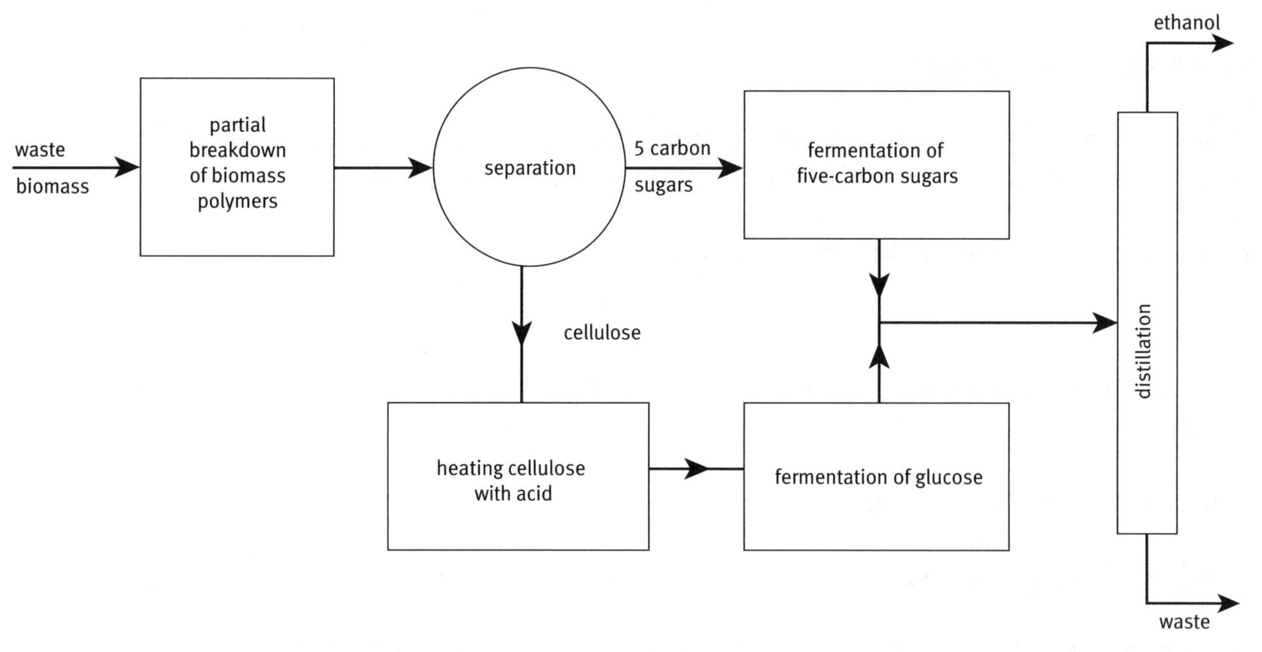

b What is the purpose of heating the cellulose with acid? ....

....

c Suggest a reason for fermenting glucose in a separate tank from xylose and other five-carbon sugars.

....

....

d What is the purpose of the distillation? ....

....

46 Uses of ethanol

Give examples to illustrate these uses of ethanol that is made on an industrial scale:

- as a fuel
- as a solvent
- as a feedstock for other processes

---

47 Ethanol manufacture and the environment

The diagram below compares, in outline, fuel ethanol from crops with fuel ethanol from oil. Crop-based ethanol could be carbon neutral.

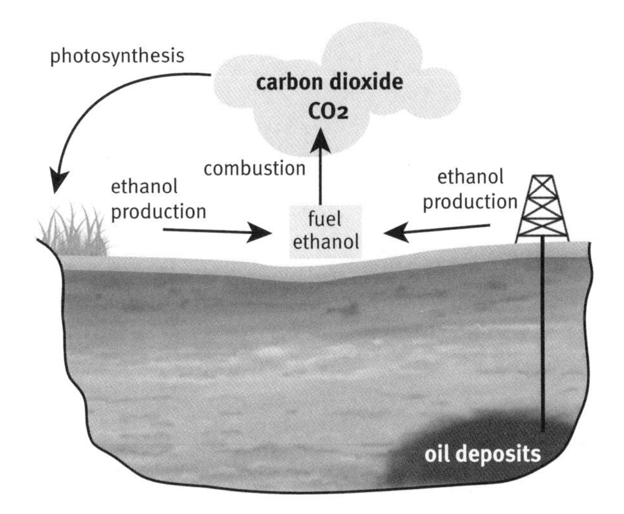

a Explain what the term 'carbon neutral' means.

b Why is fuel ethanol made from oil not carbon neutral?

c Why is it desirable to use fuels that are carbon neutral?

d Suggest two reasons why crop-based ethanol cannot be carbon neutral in practice.

e Identify two disadvantages of growing crops to make crop-based ethanol.